REPRESENTATION AND REBELLION

Representation and Rebellion

The Rockefeller Plan at the Colorado Fuel and Iron Company, 1914–1942

Jonathan H. Rees

Montante Family Library
D'Youville College

University Press of Colorado

© 2010 by the University Press of Colorado

Published by the University Press of Colorado
5589 Arapahoe Avenue, Suite 206C
Boulder, Colorado 80303

All rights reserved
Printed in the United States of America

 The University Press of Colorado is a proud member of the Association of American University Presses.

The University Press of Colorado is a cooperative publishing enterprise supported, in part, by Adams State College, Colorado State University, Fort Lewis College, Mesa State College, Metropolitan State College of Denver, University of Colorado, University of Northern Colorado, and Western State College of Colorado.

∞ The paper used in this publication meets the minimum requirements of the American National Standard for Information Sciences—Permanence of Paper for Printed Library Materials. ANSI Z39.48-1992

Library of Congress Cataloging-in-Publication Data

Rees, Jonathan, 1966–
  Representation and rebellion : the Rockefeller plan at the Colorado Fuel and Iron Company, 1914–1942 / Jonathan H. Rees.
    p. cm.
  Includes bibliographical references and index.
  ISBN 978-0-87081-964-3 (hbk. : alk. paper) 1. Colorado Fuel and Iron Company—Management—History. 2. Steel industry and trade—Management—Employee participation—Colorado—History—20th century. 3. Industrial relations—Colorado—History—20th century. I. Title.
  HD9519.C58R44 2009
  331.88'1223340978809041—dc22

                                                                    2009037502

Design by Daniel Pratt

18 17 16 15 14 13 12 11 10 09   10 9 8 7 6 5 4 3 2 1

To Laura, with love.

CONTENTS

List of Illustrations	ix
Preface	xi
Acknowledgments	xix
INTRODUCTION	1
CHAPTER ONE: Memories of a Massacre	13
CHAPTER TWO: Student and Teacher	37
CHAPTER THREE: Between Two Extremes	61
CHAPTER FOUR: Divisions in the Ranks	85
CHAPTER FIVE: The Rockefeller Plan in Action: The Mines	111
CHAPTER SIX: The Rockefeller Plan in Action: The Mill	135
CHAPTER SEVEN: New Union, Same Struggle	159
CHAPTER EIGHT: Depression, Frustration, and Real Competition	181
CONCLUSION	207
APPENDIX 1. *The Colorado Industrial Plan (also known as the Rockefeller Plan) and the Memorandum of Agreement*	221
APPENDIX 2. *Employee Representatives at CF&I Coal Mines, 1915–1928, and Employee Representatives at Minnequa Works, 1916–1928*	239
Notes	267
Bibliographic Essay	313
Index	317

ILLUSTRATIONS

FIGURES

P.1. Map of southern and western Colorado coal fields, ca. 1920	xii
1.1. Eleven-year-old Frank Snyder, shot in the melee at Ludlow during the massacre in 1914	17
1.2. Mackenzie King and John D. Rockefeller Jr. don miner's garb during Rockefeller's first tour of Colorado, 1915	31
1.3. John D. Rockefeller Jr. surrounded by the first group of employee representatives during his visit to Pueblo, October 1915	34
3.1. Diagram of the grievance procedure laid out under the Rockefeller Plan	66
4.1. Andrew J. Diamond	87
4.2. Float from a CF&I "Field Day" parade, 1920	91
4.3. Personnel card of Valentio (or Valente) Martinez, who was killed in an accident at CF&I's Minnequa Works in 1927	98
5.1. Employee representatives, managers, and their sons from Berwind at a dinner in 1923	116
6.1. Former CF&I steelworker Earl Ostrander, ca. 1929	148
8.1. The Steel Works YMCA, Pueblo, CO	188
8.2. National Labor Relations Board election, 1942	204

ILLUSTRATIONS

TABLES

4.1. Turnover at Minnequa Works, 1922–1929 94
4.2. Newly hired Mexicans at Minnequa Works, 1916–1920 99
8.1. Employee representatives at Minnequa Works, 1938 197
A2.1. Employee representatives at CF&I coal mines, 1915–1928 239
A2.2. Employee representatives at Minnequa Works, 1916–1928 254

PREFACE

> Saints have their crosses, the Rockefellers have Colorado Fuel & Iron.
> —"ROCKEFELLER'S CROSS," *TIME MAGAZINE*, AUGUST 14, 1933[1]

The Colorado Fuel and Iron Company (CF&I) came into existence as a result of the merger of the Colorado Coal and Iron Company and the Colorado Fuel Company in 1892. By 1910 it employed approximately 15,000 people, about one-tenth of Colorado's entire workforce. By the 1920s it was the largest industrial corporation in the state. Its mines produced 30 percent of Colorado's coal, while its Minnequa Works in Pueblo produced about 2 percent of all U.S. steel. CF&I owned two lime quarries, a calcite mine (both substances necessary for making steel), and iron ore mines in New Mexico and Sunrise, Wyoming. CF&I also owned a railroad, the Colorado and Wyoming, which transported elements needed for production to the company's far-flung facilities and its finished products to market. In short, the company was a prototypical vertically integrated corporation. CF&I's preeminent historian, H. Lee Scamehorn, has written that the company "played a key role in shaping the history of Colorado and the Rocky Mountain region."[2]

Like many other American steel and mining concerns, Colorado Fuel and Iron struggled during the late twentieth century. Its last mines closed in the early 1980s. The remaining steelmaking operation suffered

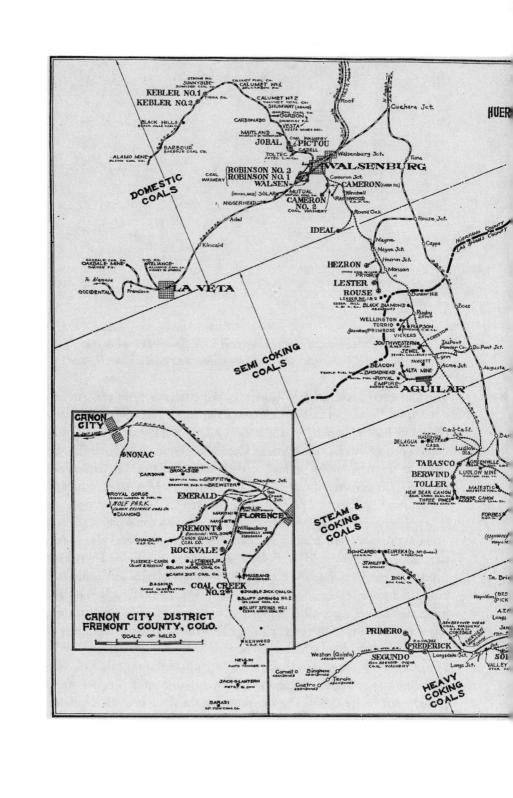

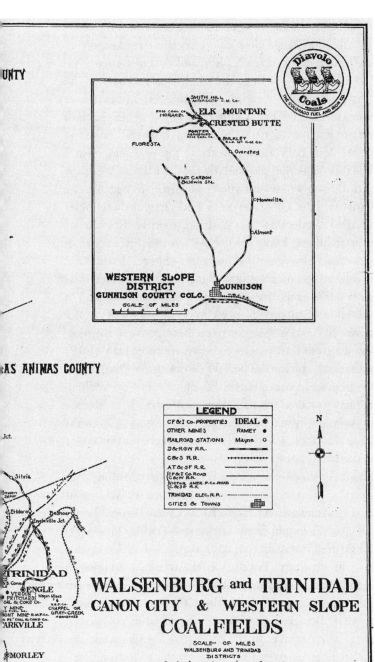

P.1. *Map of southern and western Colorado coal fields, ca. 1920. Courtesy, Bessemer Historical Society.*

greatly from the increase in imported steel that began during the 1960s and accelerated in subsequent decades. CF&I declared bankruptcy in 1993.[3] Many of the firm's records were simply abandoned in a circa 1900 office complex across Interstate 25 from the steelworks in Pueblo. The Colorado Fuel and Iron Archives contains approximately 21,000 linear feet of records related to CF&I. That is larger than most major archives devoted to collections that cover many topics (such as the Colorado Historical Society), let alone collections devoted to just one company.

Yet what makes this collection particularly special is that CF&I was not just any company. It is probably best known to historians because the Rockefeller family (of Standard Oil fame) was CF&I's largest shareholder between 1903 and 1945. Reflecting that position, John D. Rockefeller Jr. (the oil baron's first child and only son) served on the firm's board of directors. Because Rockefeller represented his family's interest in the firm, he played a major role in determining how the company conducted its business. When many CF&I employees and their family members died in the infamous 1914 Ludlow Massacre, the association with the Rockefeller name kept the incident in the headlines for years. In response to the massacre, the company instituted an employee representation plan (ERP) among its workers that was quickly dubbed the "Rockefeller Plan," since Rockefeller was the plan's chief proponent. The plan's author, William Lyon Mackenzie King, a former Canadian minister of labor, later became the longest-serving prime minister in that country's history. Countless other firms would use the Rockefeller Plan as a starting point for similar arrangements over the next twenty-plus years.

I first wrote about the Rockefeller Plan in my dissertation on labor relations in the American steel industry between 1892 and 1937.[4] In that work I offered a cursory examination of the ERP at CF&I. When I began teaching at what is now Colorado State University–Pueblo, located in the town where CF&I had its main base of operations, I had no plans to revisit that subject in any form.[5] Yet the CF&I Archives contained so much interesting and previously unseen material on the Rockefeller Plan that I decided to revisit the plan in depth. In 2000 I became a founding board member of the Bessemer Historical Society,[6] which now owns and administers the CF&I Archives in the company's Pueblo headquarters complex.[7] I am no longer involved in managing the collection, but I have advised archivists working there and have been fortunate enough

PREFACE

to get updates when they found one set of remarkable documents after another.⁸

The accounts of the Rockefeller Plan cited in the work I did prior to moving to Pueblo were all from outside observers. Therefore, I wondered what the workers themselves thought about the plan. How did they really feel about an arrangement supposedly created on their behalf? How did the plan operate on a day-to-day basis? I particularly wanted to find the minutes from meetings convened under the Rockefeller Plan in the archives. Those of us working on this project found records of a few individual grievances, but they were insufficient to allow a general assessment of the plan. Then, in August 2005, Bessemer Historical Society archivist Jay Trask informed me that he and Greg Patmore from the University of Sydney (Australia) had discovered a large cache of minutes from the joint employer/employee meetings, extending from the plan's early history to the late 1920s. The archivists later found a few sets of minutes from annual meetings at the steelworks during the 1930s.⁹ Since my question regarding what workers thought of the plan could now be answered with some certainty, I decided to write this book.

This study explains the context and structure of the Rockefeller Plan in depth and details the history of the plan in action. The records in the CF&I Archives offer not only an unprecedented view of the internal operation of a so-called company union but also an extraordinarily rare opportunity to examine the voices of non-union workers largely in their own words. Nevertheless, the study addresses both this shop-floor perspective and the view of the plan from the highest ranks of management. Strangely, regardless of one's perspective, the impact of the plan on workers' lives remains clear. Almost from the moment I started reading the records, I began to see the organization in a new light. The Rockefeller Plan does not fit the stereotype of an employer-dominated company union. While it appears that workers sometimes told management what they knew the employer wanted to hear, employees also used the plan in countless instances to complain vociferously about the terms and conditions of their employment. These workers could have been fired for speaking their minds. The fact that they did so anyway and were not discharged suggests that some CF&I workers tried to make industrial democracy work for them. In fact, industrial democracy worked very well for some employees at CF&I, albeit often at the expense of other employees.

Many other CF&I workers chose not to utilize the plan or speak their minds to management.

In short, the story of the Rockefeller Plan does not fit attacks from the left that denounce such arrangements, nor does it fit the kind of praise Rockefeller's comparatively benevolent action in setting up this organization often elicits from the right. The total story is much more complicated, and the existence of the CF&I Archives allows that story to be told here for the first time.

Because of Rockefeller's direct involvement in the operation of the plan and because of his elaborate philosophy (worked out in conjunction with Mackenzie King), I see the plan not as a successful attempt to thwart unionization but as an unsuccessful attempt to solve what was known in that era as the "labor question"—namely, how to get workers to accept the difficult circumstances created by industrialization without striking or otherwise disrupting the era's usually burgeoning economy. Americans from all walks of life were concerned with the labor question during the late nineteenth and early twentieth centuries, but managers of companies in labor-intensive industries were undoubtedly the most interested. After all, they depended on maintaining a productive workforce to keep their businesses profitable. Yet a productive workforce did not necessarily mean a happy workforce. Many employers were willing to treat their workers as disposable assets, easily replaced by others more desperate no matter what effect these policies had on employee morale. However, a few visionaries thought treating workers better might pay benefits in the long run. One of these was John D. Rockefeller Jr. He was not the most radical labor reformer of his era, but his prominence as a member of the Rockefeller family made him one of the most important.

I believe Rockefeller is best understood as coming from the paternalist tradition of the early proprietors in Lowell, Massachusetts, or the Pullman Palace Car Company magnate George Pullman rather than comparing him with other industrialists of his age, including his own father. The younger Rockefeller set his entire family on a different course from that of other extremely wealthy families of the era. This explains why, despite the terrible press immediately following the Ludlow Massacre, William Manchester could write in 1959 that "[t]he Rockefellers became rich under John D., but popular under Junior."[10] Nothing underscores the reason for this distinction more than the different ways the two men treated the Colorado Fuel and Iron Company. After the massacre, John D. Rockefeller Jr. tried to create a new labor policy that would be both humane and profitable for shareholders like him. In contrast, John D. Rockefeller Sr. not only sold all his personal shares in CF&I at that time but sold them short.[11] No wonder the senior Rockefeller was the richest man in the world while his son suffered from indigestion and migraine headaches his whole life.

ACKNOWLEDGMENTS

Thanks to the archivists at the Bessemer Historical Society who were working there during the time of my research for this book: Jay Trask, Bev Allen, Wendy Fairchild, and M'lissa Morgan. They made their reading room a home away from home. Greg Patmore started studying CF&I long before this book project began and has been a great help in placing the situation the company faced in terms of industrial relations within both the U.S. and world contexts. Without the Bessemer Historical Society this book would never have been written because there would have been no CF&I Archives available. Anyone who cares about industrial history should thank the society's board of directors and membership for working to save this priceless collection.

Sandy Hudock is the best interlibrary loan librarian a historian working at a school with a small library could hope to have. The students in my special topics class "Mine and Mill" during the Spring 2007 semester at CSU-Pueblo were a great sounding board for some of my thoughts about CF&I and produced solid research of their own about the company.

Amy Fitch at the Rockefeller Archive Center, Eric Bittner at the National Archives Rocky Mountain Region in Lakewood, Colorado, Karen Kukil at Smith College, and David Hays at the University Archives at the University of Colorado–Boulder were a great help in finding supporting material to tell the story of the ERP. The Rockefeller Archive

Center gave me much-needed financial assistance with a travel grant. Jeffrey DeHerrera did much of the initial work needed to make the tables in the appendix possible. Jim Hawthorne helped immeasurably with his copyediting skills.

I received excellent research tips and advice from Tom Noel, John Cooper, Bruce Kaufman, Joe McCartin, Ray Hogler, Fawn-Amber Montoya, Bob Reynolds, Rogers Hollingsworth, the late Robert Feinman, the Business and Labour History Group at the University of Sydney (Australia), and especially Richard Myers. Special thanks to Thomas Andrews for suggesting the title. My first objective when I started this project was to use the CF&I Archives merely to create a journal article suggesting that the Rockefeller Plan was more complicated than its reputation among historians suggests. Gerald Friedman and Craig Phelan, the editors of *Labor History*, were kind enough to publish that piece. Having now done much more research on this subject, I would change many aspects of that article. Nevertheless, it is reassuring to know that the essential argument about the plan's benefits and limitations for workers still holds up in the face of the additional evidence I have now seen. Thanks also to the Baker Library of the Harvard Business School for permission to cite material from the Elton Mayo Papers.

REPRESENTATION AND REBELLION

INTRODUCTION

> Herein is the fundamental difference between the Rockefeller plan and that of trade unions. One develops independence, the other relies on the graciousness and good will of the employer. There can be no compromise between the two theories, for if working men are to be really free, their right to regulate their own lives must be acknowledged.
>
> FRANK MORRISON, SECRETARY OF THE AMERICAN FEDERATION OF LABOR[1]

In 1926, Ernest Richmond Burton "broadly defined" employee representation as "any established arrangement whereby the working force of a business concern is represented by persons recognized by both the management and the employees as spokesmen for the latter in conferences on matters of mutual interest."[2] At that point employee representation plans (ERPs), or company unions as their critics called them, had been present in the United States for about two decades. The term "company union" has persisted to this day as a way to describe all non-union employee representation arrangements. It strongly suggests a quality present in many ERPs: management's dominance in the operation of the plan. However, the term can be highly misleading because not all ERPs were autocratic.[3] Some observers cite Rockefeller's decision to champion the Rockefeller Plan, the employee representation plan at the Colorado Fuel and Iron Company (CF&I), as a significant reform instituted as penance for condoning the attitudes on the ground in Colorado that led to the tragic Ludlow Massacre. Labor historians, on the other hand, have tended to suggest that the plan was really an iron fist in a velvet glove, a socially acceptable way for John D. Rockefeller Jr. to keep the United Mine Workers of America (UMWA) away from CF&I yet still improve his and his company's reputation among workers and the general public.

Both positions are to some extent accurate. The tragic 1914 Ludlow Massacre led Rockefeller to become a liberal anti-unionist. Rather than fight independent trade unions with guns or injunctions, he chose to fight them with kindness. Rockefeller wanted his union to make the UMWA obsolete by doing what it did more effectively. Designed by Mackenzie King, the future prime minister of Canada, the plan offered CF&I workers rights they would not have had otherwise. It also became the vehicle for an elaborate welfare capitalism program. Unfortunately for Rockefeller, giving workers the carrot rather than the stick still failed to placate them in the long run, which says a great deal about the limits to what any ERP can achieve. Although they had substantially less power than management did, CF&I workers were still able to carve out useful benefits for themselves through the bargaining process the Rockefeller Plan created—many of which management had not originally intended to grant. Yet despite this success, most CF&I workers ultimately preferred an independent trade union to the kind of circumscribed bargaining the plan provided.

By creating an ERP, Rockefeller aspired to answer the "labor question," namely, how to get workers to accept the difficult circumstances created by industrialization. However, as was the case with many other industrial relations visionaries (particularly the anti-union ones), his effort did not bear fruit. Instead, workers in both the company's coal fields and the steel mill chafed under the restrictions management placed on its labor organization. As a result, both sides of the business eventually opted for independent unions, and the ERP has largely been forgotten. In its day the Rockefeller Plan brought employee representation plans into public consciousness for the first time because of continued public interest in Ludlow. After CF&I introduced this arrangement in 1915, similar ERPs spread rapidly during World War I and reached all corners of American industry during the 1920s. These arrangements attained their greatest popularity during the New Deal years, when many companies used them to fight legally mandated collective bargaining with independent trade unions. The Rockefeller Plan also had considerable influence in Canada, and its effects can still be seen in industrial relations practices in the United Kingdom, Japan, Germany, Australia, and (despite the prohibition on employer-dominated labor organizations contained in the 1935 National Labor Relations Act) the United States.[4]

Both sides of the labor-management divide can learn a great deal from the history of the Rockefeller Plan. Recently discovered documents from the Colorado Fuel and Iron Archives that describe the day-to-day operation of the Rockefeller Plan make it possible for the first time to closely examine and evaluate the relationship between labor and management during the ERP's existence. While trade unionists and businesspeople alike had preconceptions about what the Rockefeller Plan could or could not do, only a handful of outsiders ever saw it in operation. With the availability of documents detailing that operation, especially minutes of joint labor-management meetings, the words and actions of employee representatives and rank-and-file workers can be examined to see what they really thought of the arrangement.

Even a cursory look at the material in the CF&I Archives demonstrates that workers reacted to the plan in ways its originators did not anticipate. Many workers, especially skilled ones, appreciated the opportunity the Rockefeller Plan gave them to communicate their feelings about the terms and conditions of their employment and to benefit from management's largesse. Many other workers, particularly less-skilled employees who were often immigrants and nonwhite, recognized that the plan offered them little or nothing and ignored it entirely. Some workers expressed appreciation for the benefits of the plan but hostility toward management because its implementation limited their ability to express themselves.

The day before its introduction, Rockefeller suggested that the plan would "make strikes unnecessary and impossible."[5] He also repeatedly claimed that treating workers humanely would prove profitable for employers in the long run. These efforts to obtain stability and profitability failed miserably. Indeed, the failure to achieve stability contributed greatly to the failure to achieve profitability. Even a comparatively benevolent employee representation plan such as the Rockefeller Plan did not solve the company's labor problems because the plan's limitations on freedom of action ultimately alienated even the most loyal employees. These tensions hurt the bottom line of a company that would have faced financial difficulties even if it had achieved a peaceful labor situation. Therefore, they help explain why CF&I entered receivership in 1933 and underwent bankruptcy reorganization in 1936. The hostile response to Rockefeller's employee representation plan likely played a major role in destroying the company's financial position. It is only fitting then that the Rockefeller family held the majority of the company's stock when bankruptcy occurred.

THE "HUMAN ELEMENT" IN INDUSTRY

Sometime around 1920, Colorado Fuel and Iron published two cards printed on colored cardboard. One, "'The Present Need and the New Emphasis Within Industry' or the Industrial Representation Plan and Leadership," was intended for foremen and superintendents. The other, "The Purpose and Principles of the Industrial Representation Plan and the Best Methods of Securing Maximum Benefit from Its Administration," was intended for employee representatives, the workers elected to represent labor's interests when bargaining with management. The company printed the cards to educate key participants in the plan about how they should carry out their duties. In both cases, these instructions reflected the difficulties management faced in testing the ideas behind the Rockefeller Plan. Indeed, the content of the cards reflects timeless complexities faced by anyone interested in influencing human behavior. The failure of the plan to have adequately addressed these complexities to that point explains why management needed to print them.

The card designed for employee representatives included a line about the underlying principle of the ERP: "The Industrial Representation Plan is based on the belief that the interests of the employee and the employer, as well as [those of] the public they both serve, are mutual and not antagonistic, and that these interests can best be furthered by co-operative effort. It assumes that every man desires only justice for himself and is equally willing that justice be done [for] all other parties."[6] But how could management convince workers that the plan was sincere in its call for cooperation? That is why the foremen and superintendents also needed instructions: they were the ones who most often dealt with employee representatives. If low-level managers failed to treat workers on the other side of the bargaining table with respect, the employee representatives would not take the plan seriously. "Management constantly faces a two-fold problem," the other card explained. "First, that of Production; second, that of wisely directing the Human Element. The outstanding present need among men within industry is that of a real spirit of democracy actually at work all the time."[7] These sentiments embodied John D. Rockefeller Jr.'s philosophy of labor-management relations. "One of the reasons there are so many labor troubles is that we have forgotten the human element," Rockefeller told an interviewer in 1917. "Labor is being looked upon as a commodity—as part of an equipment—as something that may be bought and sold. We sometimes forget that we're dealing with human beings. . . . The big thing we've got to do is inject the spirit of brotherhood into the labor question. There is no other way."[8] For Rockefeller, employee representation was the best vehicle for bringing this spirit of brotherhood into industrial relations.

Rockefeller's philosophy worked less well in practice than it did in principle. For one thing, even if the Rockefeller Plan were a viable method for workers to communicate their problems to management, management could not force workers to use it or even to be honest with their bosses. Workers generally recognized that the power management held over them might adversely affect them if they made use of this arrangement. "Take it up with the representatives you say[?]" asked the chair of the strike committee at the steelworks during the 1919 dispute there. "It is just like bringing suit against the devil and holding court in hell."[9] Ben Selekman, who interviewed miners that same year, reported, "On my various visits to the mines, I have always mixed with the various groups of men, and attempted to sound them out on the R. Plan. They invariably say, 'To hell with the R. Plan, it is no good, it is one sided.'"[10] But this does not mean the Rockefeller Plan offered labor nothing of value.

CF&I wanted the plan to provide stability in its labor relations. Labor historians such as David Montgomery, for example, have covered the potential benefits of the Rockefeller Plan to management (assuming it worked as intended) but say little or nothing about the potential benefits to employees.[11] Workers and especially employee representatives at CF&I used both the power they had on the shop floor and management's paternalistic attitude toward them to gain many improvements through the

ERP. To automatically dismiss such victories as somehow tainted because these people chose to work within a flawed system does not do justice to the achievements.

Likewise, to automatically dismiss Rockefeller's ideas about workplace democracy as the self-serving rhetoric of a man who could afford to be generous would also be a mistake. Considerable evidence attests to Rockefeller's sincerity. As Bruce Kaufman has explained, he pressured other companies connected to his family to liberalize their industrial relations policies, supported employee representation at President Woodrow Wilson's Industrial Conference in 1919 despite opposition from other businesspeople, and created Industrial Relations Counselors, Inc., to help employers construct labor policies that reflected his humanistic views on labor issues.[12] Rockefeller also gave numerous public speeches detailing an intricate philosophy of industrial relations based on the ideas of Mackenzie King. Indeed, Rockefeller not only talked the talk of labor reform following the Ludlow Massacre, he lived the life of a Christian liberal. Under his direction the Rockefeller family gave millions of dollars to progressive ecumenical groups such as the YMCA and the Interchurch World Movement, and Rockefeller tried to influence those groups to adopt more modern stances on numerous economic issues.[13]

By the late 1910s, John D. Rockefeller Jr. had become the spokesperson among American businesspeople for liberal anti-unionism. As Howard Gitelman, no friend of Rockefeller, has explained:

> Simply by continuing to advocate employee representation . . . Rockefeller came to appear more and more progressive. At first, this feat of seeming to move while actually standing still was an illusion, a mirage created by the chaotic rightward drift of a nation on the verge of panic. In time, however, as the political spectrum of the country shrank to the narrowest band of tolerance, the illusion became increasingly real. In a business environment presided over by the great steel industry [of] Babbitt, Elbert H. Gary, Rockefeller *was* something of a very modest liberal.[14]

In fact, Gary Dean Best has argued that the movement for employee representation Rockefeller championed "can be legitimately regarded as a part of the Progressive movement."[15] So while labor historians tend to paint the worst possible portrait of the industrialist, the way one assesses John D. Rockefeller Jr. actually depends upon one's point of comparison.

INTRODUCTION

While Rockefeller appears reactionary when compared with the leaders of unionized coal firms in the East, he seems liberal next to U.S. Steel's Elbert Gary. The only thing the workers cared about, however, was what the Rockefeller Plan offered them. Because of the great damage to his family's reputation as a result of Ludlow, the plan offered workers a great deal. While CF&I employees might have done better had they joined independent unions, many actively participated in the Rockefeller Plan because it was the best option available. Both the company's coal miners and steelworkers were better off working under the Rockefeller Plan than they would have been working under no union at all.

Nevertheless, CF&I employees had to participate in the plan in order to benefit from it. Unfortunately for management, many rank-and-file CF&I workers at both the mines and the mill chose not to take part in the plan or never even knew about it. Many employees objected to steps management took to prevent genuine unions from emerging. In fact, no matter how beneficial or altruistic any management-dominated union might have been, ERPs such as the one at CF&I inevitably paled in comparison to independent trade unions from the workers' perspective because they could not offer workers the freedom to act independent of management. Whenever employee representatives at CF&I tested management's willingness to make concessions, their failures reminded them that real trade unions imposed no such restrictions, thereby whetting their desire to join independent organizations down the line. This explains why anti-union stalwarts such as Elbert Gary opposed employee representation in all circumstances.[16] Creating an employee representation plan was a sign of weakness. Usually, companies that bargained with such organizations were compelled to do so either by the federal government or by the success of an outside union trying to organize them. Executives like Gary were unwilling to risk having a pseudo-union turn into the real thing. From Gary's standpoint, Rockefeller's willingness to experiment with ERPs was therefore dangerously radical.

Not coincidentally, independent trade unions opposed ERPs for reasons opposite Gary's. It is ironic then that the person who most effectively argued that "company unions" such as the Rockefeller Plan could actually help independent trade unions was American Federation of Labor (AFL) president Samuel Gompers. As Gompers argued shortly after the Rockefeller Plan debuted in 1915, "[T]he Rockefeller unions will help—

they will be an educational opportunity. Whenever men meet together to talk over working conditions and ways of betterment, there enters into their lives an incalculable opportunity for progress. Mr. Rockefeller is laying the foundations upon which real unions will be developed."[17] While Gompers recognized that the Rockefeller Plan was an anti-union device, he thought management's failure to consider the sense of independence and dignity that membership in an independent trade union provided would eventually bring CF&I workers over to those unions. Gompers changed his mind about employee representation plans in 1919 when workers trying to organize the steel industry introduced a resolution against them at the AFL convention, but his first instinct ultimately proved correct, at least in the case of the Rockefeller Plan.

"THEY ARE BETTER BARGAINERS THAN WE ARE"

When the National Industrial Recovery Act (NIRA) passed in 1933, many employers throughout the United States scrambled to establish ERPs to block their workers from joining independent unions. The language in Section 7(a) of that law gave workers the right to join unions of their own choosing and explicitly protected them from having to join employer-dominated organizations. Apparently democratic employee representation plans gave bosses a way to follow the letter of the law without incurring the perceived drawbacks of recognizing independent unions in their shops. Often merely union-avoidance strategies, the growth of these ERPs for this purpose at this time reinforced opposition to all "company unions" within labor movement circles.[18] When New York senator Robert Wagner offered the bill that would become the National Labor Relations Act (NLRA) in 1935, he included a similar provision to the one in NIRA Section 7(a) that outlawed management-dominated unions. Employers generally waited until the Supreme Court's Jones and Laughlin decision upheld the constitutionality of the NLRA before abandoning these anti-union efforts.[19] That decision marked the end of "company unions" in the traditional sense of the term. The labor movement has not missed these organizations.

For the most part, labor historians have taken up where the unions left off, condemning the employee representation plans of the NLRA era in part to oppose modern legislation that threatened to loosen restrictions on such organizations. Bruce Kaufman has organized the chorus of academic criticism of ERPs into four arguments: they were created as union-avoidance strategies, they were "sham" organizations (meaning workers had little power), they could not protect employee interests, and they were significant barriers to the growth and expansion of trade unionism.[20] As a result of such characterizations, ERPs have received little coverage from labor historians compared with their historical significance as an anti-union tool.[21] While all these criticisms are valid with respect to many ERPs, particularly those instituted during the 1930s, they fit the Rockefeller Plan less well. In fact, this academic hostility toward "company unions" has largely hidden the fact that not all ERPs were the same.

The history of the Rockefeller Plan demonstrates that employee organizations do not have to be free from employer influence to improve workers' lives. As the Federal Council of Churches explained in its study of the 1927–1928 Colorado mine strike, "It is generally conceded even by critics that a great improvement has been made in living and working conditions [at CF&I] during the life of the plan."[22] Nevertheless, workers' attitudes toward the plan depended upon each individual's particular situation. Independent trade unions, while generally good for workers who could join them, did not necessarily act in the best interests of non-union workers such as those at CF&I. The survival of independent unions depended upon workers demanding representation by them rather than accepting half-measures like an ERP. Many non-union workers actually appreciated the improvements their bosses were willing to make in the terms and conditions of their employment through ERPs. John R. Commons compared the AFL's eventual position on employee representation plans to "revolutionary socialism." He thought labor leaders were in effect saying, "It is better to let conditions get as bad as possible because only then is revolution attractive to the oppressed."[23]

Some miners and steelworkers at CF&I had little desire to wait for the entire industrial relations system to change so their working lives could improve. As one employee representative at a Fremont County mine explained in 1927 (during a meeting he did not know was being transcribed), "I have been an officer in the United Mine Workers, but when the Industrial Plan came along and gave us practically what we were asking for through that organization, I was satisfied with it as a scientific

INTRODUCTION

way to bargain collectively."[24] Even though it took many years for the Rockefeller Plan to become an independent union in both segments of CF&I's business, miners and steelworkers used the voice it gave them to significantly improve the terms and conditions of their employment in the interim. From the workers' point of view, immediate half-measures were better than accepting nothing from their employer until the time for independent unions arrived.

While those who agree with the labor movement's traditional opposition to ERPs might see the cost of these improvements as a small price for management to pay in exchange for a pliant workforce, they should recognize that CF&I miners and steelworkers remained remarkably militant during the years the Rockefeller Plan was in effect. In fact, the Rockefeller Plan did not end labor-management tension at CF&I; it merely redirected much of it into a structured format. In the coal mines, these tensions frequently spilled over into strikes for organization. Even in the steel mill, whose workers had no viable outside union to join, employee representatives consistently pressed demands on management in the strongest possible terms. These were not the management "stooges" who gave employee representation plans a bad name during the 1930s. CF&I employees did their best to take the structure management imposed upon them and make it operate like the independent outside unions they did not have. In response to these efforts, management continually made concessions in both the mines and the mill to keep workers happy. "They are better bargainers than we are," explained steelworks manager Louis F. Quigg to the National Labor Relations Board in 1938. "They were always better bargainers, and that is what they wanted."[25] The chapters in this book present abundant evidence that Quigg's assessment was correct.

The circumstances that led Colorado Fuel and Iron to implement the Rockefeller Plan, the philosophy behind it, and the specific details of the plan are very important for understanding its successes and failures. Chapter 1 discusses the firestorm of public criticism that descended upon John D. Rockefeller Jr. following the Ludlow Massacre, specifically the withering assault on his reputation led by the United Mine Workers of America and its supporters. Rockefeller's desire to prevent future criticism largely explains why CF&I was so much more accommodating to workers than were other employers who created employee representation plans after CF&I did so. Chapter 2 covers John D. Rockefeller Jr. and Mackenzie King's philosophy of industrial relations. Rockefeller's belief in the mutual interests of labor and management (which he learned from King) separated him not only from the leaders of independent labor unions but also from most other industrialists of the day who never considered treating labor as anything but a commodity. This belief also explains management's willingness to make many concessions to employees under the auspices of the plan. Yet while the two men eventually agreed on almost everything, King never convinced Rockefeller to stop treating employee representation as a substitute for independent unions. Chapter 3 closely examines the structure of the plan, particularly those provisions that were rare or unique in company-initiated employee representation plans of the era, such as the outside arbitration clause. The chapter also highlights the close relationship between the operation of the ERP and CF&I's pioneering welfare capitalism program.

Chapter 4 discusses the response to the plan among different groups of CF&I employees. Racial and ethnic tensions in the mines and mill help explain why some CF&I employees responded differently to the ERP than others did. Skilled white workers used the plan the most in both the mines and the mill because they had the tenure and influence to win positions as employee representatives year after year. Unfortunately, they used the positions to benefit those like them rather than to help the entire workforce, thereby fomenting the kind of unhappiness the plan had been created to prevent. Likewise, different conditions of employment in the steel and coal mining industries explain the relative willingness of aggrieved CF&I workers to express their concerns on the job or on the picket line, as well as their success at getting management to change those conditions. Chapter 5 discusses how the plan played out in the mines, based primarily on the minutes of meetings held while the Rockefeller Plan was in force, in both calm periods and the frequent strikes during its history. Chapter 6 considers the operation of the plan in the steel mill. Many of the differences between the way the plan operated there and how it operated in the coal mines stem from the absence of a viable independent union like the UMWA that steelworkers could aspire to join. Yet while steelworkers walked off the job only once during the existence of the

Rockefeller Plan at the Minnequa Works, steelworkers were even more militant than the miners in using the ERP as a vehicle to make demands on management.

The rapid decline of the plan from experiment in corporate liberalism to typical anti-union device began during the 1927–1928 Colorado mine strike, which was led by the Industrial Workers of the World (IWW). Chapter 7 examines how the IWW used the Rockefeller Plan as an organizing tool prior to its 1927–1928 strike and explains the relationship between the plan and the successful organizing drive by the UMWA that led to the arrangement being eliminated from CF&I mines in 1933. Chapter 8 considers the operation of the plan in the steel mill during the 1930s and the struggle by the Steel Workers Organizing Committee to organize the mill both on the shop floor and in court. The Conclusion looks at the way the story of the Rockefeller Plan relates to the debate over the "company union" clause in the National Labor Relations Act, an important subject in industrial relations over the past twenty years. Taken as a whole, the history of the Rockefeller Plan is not the story of ceaseless oppression and stifled militancy its critics might imagine, but it is also not the story of the paternalist panacea for labor unrest John D. Rockefeller Jr. hoped it would be.

Chapter One

MEMORIES OF A MASSACRE

> What can businessmen do to clean up the rot that these muckrakers and demagogues have dumped on our door?
>
> —COLORADO FUEL AND IRON COMPANY CHAIR AND VICE PRESIDENT LAMONT BOWERS, 1914[1]

In 1918, John D. Rockefeller Jr. made his second trip to Colorado since the Great Coalfield War of 1913–1914. On May 30 a chauffeur-driven car carrying him, his wife, Abby, and Mackenzie King arrived at a gathering of approximately 3,000 working people in southern Colorado. This multiethnic and multiracial crowd had assembled for the dedication of a monument to the victims of the Ludlow Massacre, which had occurred at that spot during the infamous strike a little more than four years earlier. A few of the leaders of the United Mine Workers of America (UMWA) who had organized the dedication ceremony for the monument had learned the previous evening that Rockefeller planned on attending and wanted to speak, but they had not yet decided on a response. Rockefeller's presence at the gathering was an issue because many in the audience felt he was responsible for the deaths of the people being memorialized. In his role as primary stockholder, director, and de facto owner of the Colorado Fuel and Iron Company (CF&I)—the victims' employer—Rockefeller had supported management's uncompromising refusal to bargain with the union during the coal field war. Union leaders were afraid there would be an embarrassing or even a dangerous incident if he attended, let alone spoke. They communicated this fear to King, who went back into the vehicle to explain the situation to the passengers. The Rockefellers sped off without leaving the car.[2]

Why did Rockefeller want to speak at the dedication of the Ludlow memorial? Since he did not accept responsibility for the Ludlow Massacre or even acknowledge that it had occurred, guilt was not the reason. "There was no Ludlow Massacre," he wrote to his public relations specialist Ivy Lee shortly after the incident. "The engagement started as a desperate fight for life by two small squads of militia, numbering twelve and twenty-two men. There were no women and children shot by the authorities of the State or representatives of the operators in connection with the Ludlow engagement. Not one."[3] While this statement was technically true, it contradicted the overwhelming sentiment of the general public at the time.[4] His willingness to repeat this same argument before the U.S. Industrial Relations Commission in 1915 demonstrated that he had failed to understand that fact. By the time Rockefeller arrived in Colorado for his second tour in 1918, he well understood the public hostility toward him and was taking steps to reverse that sentiment.

John D. Rockefeller Jr. (and, by extension, his family) had a major image problem long before gunfire broke out at the Ludlow tent colony as a result of the mine owners' poor handling of the 1913–1914 miners' strike that preceded it. Henry Demarest Lloyd's 1894 book *Wealth against Commonwealth* and especially Ida Tarbell's 1904 *History of the Standard Oil Company* had already done significant damage to the Rockefeller name.[5] Rockefeller Jr. specifically had become a flashpoint for workers prior to Ludlow. On April 6, 1914, he testified before a congressional committee investigating the strike. A member of Congress asked if he would continue to support the open shop (i.e., union-free) labor policy at CF&I's coal camps even "if it costs all your property and kills all your employees." "It [the open shop] is a great principle," he replied.[6] The quote appeared frequently in critics' attacks on Rockefeller after the Ludlow Massacre.[7] Even more ominous, the Industrial Relations Commission discovered the next year that despite public testimony to the contrary, Rockefeller kept in constant touch with CF&I management during the strike and was, to quote Chairman Frank Walsh, "the directing mind throughout the struggle."[8] Therefore, while he might not have ordered the massacre at Ludlow, Rockefeller certainly understood that sending the Colorado National Guard into the coal district made catastrophic violence a distinct possibility. He thus bears considerable responsibility for the tragedy even though he was in New York when it happened. Rockefeller also bears responsibility because of his repeated promises during the strike to back the anti-union policies of CF&I management in Colorado to whatever extreme it decided to take them. "What ever the outcome may be," Rockefeller wrote Lamont Montgomery Bowers at one point during the strike, "we will stand by you to the end."[9]

Less recognized than Rockefeller's culpability are the considerable efforts of the UMWA and its supporters to hold him personally responsible for the deaths at Ludlow. In the short run, the UMWA hoped to fan the fires of resentment through its own public relations campaign, as well as through its actions in relation to multiple government investigations of the coal strike and the massacre. In the months following the massacre, the UMWA took advantage of Rockefeller's poor grasp of public relations to turn him into a villain. Only when he announced the Rockefeller Plan did he begin to counter that impression. For this strategy to succeed in the long term, the employee representation plan (ERP) Rockefeller and King created needed to be substantive. Cosmetic industrial relations reform—or, even worse, an oppressive anti-union tool—would not have prevented future labor troubles in Colorado. Such a situation would have led to continued bad press for Rockefeller and his family, a prospect he found intolerable.

THE LUDLOW MASSACRE

Not surprisingly, the coal companies blamed the 1913–1914 strike and the resulting violence on the union. According to a report of coal mine managers (ghostwritten by Ivy Lee), "*This strike was in its inception, and always has been, a strike for union recognition only*" (emphasis in original).[10] Another bulletin from employers in July 1915 offers more detail regarding management's thinking: "This strike was not the work either of the managers of the mines or of any large portion of the miners, less than 10 per cent of whom were members of the United Mine Workers of America. It was planned outside the state of Colorado, led by outsiders, and financed from outside."[11] Unfortunately for Rockefeller, such classic anti-union propaganda was not convincing to the general public. As a result of the Ludlow Massacre, the public became highly interested in the miners' struggle for organization.

The Ludlow Massacre, by far the best-known event of the 1913–1914 strike, was one of the bloodiest events in American labor history. On the morning of April 20, 1914, gunfire broke out between striking miners and a battalion of the Colorado National Guard at the Ludlow tent colony, where the miners and their families were living after having been kicked out of company-owned houses the previous year. Nobody knows who fired the first shot, but because the Guard had machine guns, the other side took most of the casualties. Women and children fled in terror from the scene even before the Guard set fire to the tents. Eleven children and two women who were unable to escape died in a pit under one of the tents, where they had gone to avoid being in the line of fire. These shocking casualties, as well as the deaths of three union members (including the legendary Louis Tikas) after they had surrendered, are the primary reasons this battle is known as a "massacre."[12] When the striking miners learned what had happened at Ludlow, they attacked mines in coal fields all over the state, resulting in significant casualties on both sides. In response, on April 30 President Woodrow Wilson sent federal troops to Colorado to end the bloodshed.[13]

The Ludlow monument whose dedication Rockefeller had wanted to attend lists seventeen names. They include the women and children found in the "Death Pit" after the massacre, Frank Snyder (a young boy shot during the melee), and Louis Tikas and his two colleagues who were killed while under the white flag of surrender. This number is low compared to other figures suggested in the years following the event. Just two weeks before the dedication of the monument, the United Mine Workers Journal noted that "33 men, women and children were brutally slain" at Ludlow.[14] This emphasis on dead women and children accentuated the impression that union miners and their families had been victims throughout the strike, an impression that continues to this day. This impression is highly misleading.

The United Mine Workers of America also caused casualties during the Great Coalfield War. Scott Martelle has documented seventy-five deaths from the beginning of the strike through the arrival of the U.S. Army in response to post-Ludlow violence.[15] That figure includes thirty-

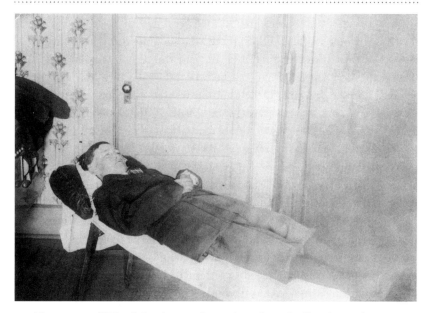

1.1. *Eleven-year-old Frank Snyder was shot in the melee at Ludlow during the massacre in 1914. Industrial Relations Commission chair Frank Walsh thrust this picture (as opposed to one of a dead adult miner) before John D. Rockefeller Jr. during his testimony to evoke sympathy for the strikers' cause. Courtesy, Western History Collection, Denver Public Library, Z-214.*

three strikebreakers, mine guards, and members of the Colorado National Guard—most of whom were probably shot by strikers or other labor sympathizers.[16] Immediately following the massacre, the local UMWA and its labor movement allies issued a call to arms. While phrased as a gathering together of arms "for defensive purposes," others saw it as an insurrection that had to be put down by the U.S. Army.[17] By Martelle's accounting, the deaths of mine guards and militia members as a result of this action surpassed those at Ludlow on both sides.[18] Despite such union-led or union-inspired violence, the press and the public's unrelenting focus on the Ludlow Massacre essentially wiped out public memory of the union's aggressive stance. In a perverted way, then, the Ludlow Massacre was a gift to the United Mine Workers of America. Because of what happened on April 20, 1914, the other acts of violence connected to the strike have been largely forgotten.

In a society that still maintained vestiges of Victorianism, the deaths of women and children were so shocking that they caused all the other casualties to be ignored. In fact, the loss of women and children at Ludlow turned what had been a lost strike into an occasion of sympathy for the union. "Little children roasted alive make a front page story," noted the labor activist Mary "Mother" Jones incorrectly in her autobiography.[19] "Dying by inches of starvation and exposure does not."[20] In the days following the tragedy, many people thought the death toll of women and children was much higher than it actually was. Nobody knew exactly how many noncombatants had died, so the massacre of these innocents drove coverage of the event.[21] As *The Rocky Mountain News* wrote, "The blood of women and children, burned and shot like rats, cries aloud from the ground. The great state of Colorado has failed them. It has betrayed them."[22] CF&I never fully recovered from the bad publicity during this period. Not content to rest on its laurels in a multiyear organizing struggle that depended substantially on public pressure to force operators to recognize the union, the UMWA helped sustain the bad press from Ludlow by directing the blame for the massacre toward CF&I's largest stockholder.

THE UMWA'S CAMPAIGN OF BLAME

The Rockefeller family first concerned itself with the affairs of the Colorado Fuel and Iron Company when financier Jay Gould brought in John D. Rockefeller Sr. to help finance a loan to the previous owner, John C. Osgood, in 1903. Shortly thereafter, John D. Rockefeller Jr. recommended that Gould and his family buy up related Colorado companies, including the Colorado and Wyoming Railway, to better protect the initial investment. The younger Rockefeller became a member of the company's board of directors that same year, representing the family's interests. By 1911, Rockefeller Sr. had retired, leaving his son in sole control of the family's far-flung business interests.[23] Many years later the younger Rockefeller called Ludlow "one of the most important things that ever happened to the family."[24] While Rockefeller undoubtedly thought the importance of the event stemmed from his recent interest in industrial relations, Daniel Okrent believes Ludlow's true importance for Rockefeller stemmed from the fact that "for the first time in his life,

Junior had taken an independent step. He was forty."[25] That step was the introduction of the Rockefeller Plan.

Considering Rockefeller's lack of remorse for the killings at Ludlow, Ludlow itself was not important to the family—but the firestorm of bad publicity that followed was. Countless public attacks on his father's business methods had made the younger Rockefeller particularly sensitive to the way the press portrayed him and his family. Rockefeller first came to public attention as the foreman of a grand jury investigating forced prostitution (better known as "white slavery") in 1910. Alarmed by what he learned while doing his civic duty, Rockefeller began to fund further investigation of this issue. The positive press he generated while doing this work dissipated during the Colorado strike.[26] In response, Rockefeller hired Ivy Lee (generally acknowledged as the father of modern public relations) for assistance. To better situate him for this task, Rockefeller appointed Lee to CF&I's board of directors. Lee's first significant task was to travel to Colorado and survey the situation on the ground. In August 1914, Lee reported back to Rockefeller: "The people of this state have been led to believe by the hostile press that you and your friends are exploiting the state. From friendly sources, I gather that opinion is still widely held."[27] This was a natural by-product of management's refusal to take the miners' demands seriously, instead blaming the entire dispute on outside agitators.

Public hostility toward management is easily understood because press coverage during the strike became even more critical after the massacre. The first reports bordered on the hysterical and eventually proved to be wildly inaccurate. An example is this excerpt from *The Telluride Daily Journal*:

> THE LEADERS OF THE STRIKING COAL MINERS HERE TODAY SAID THAT AT LEAST 50 PERSONS WERE DEAD AS A RESULT OF THE LUDLOW BATTLE. FURTHER INVESTIGATION INTO THE BATTLE ONLY INCREASES THE HORROR OF IT ALL.
>
> THE LABOR LEADERS HAVE ALREADY NAMED THE BATTLE "THE SLAUGHTER OF THE INNOCENTS." IT IS BELIEVED THAT MORE THAN TWO-THIRDS OF THOSE SLAIN WERE WOMEN AND CHILDREN.[28]

Much of the early coverage, as mentioned, focused specifically on the plight of the women and children. According to *The New York Times*:

"Women ran from the burning tents, some with their clothing afire, carrying their babies in their arms. Many, in order to save the babies at their breasts, were forced to abandon their older children to their fate."[29] Once CF&I became associated with such images, management's response invariably failed to fix the damage. As late as May 16, the *Literary Digest* was reporting "[t]hat a straightforward, accurate story of what happened in Colorado is still impossible, because of 'press censorship, interrupted communication, and lack of disinterested witnesses.'"[30] Unable to learn the truth surrounding Ludlow, the public continued to believe the worst.

Rather than take any responsibility for the tragedy, union opponents blamed the UMWA for spreading lies about the event. This is somewhat understandable, as many of the initial stories the union perpetrated were factually incorrect. For example, a scathing contemporary report by UMWA District 15 publicity director Walter Fink claimed, "There are more than fifty women and children missing and it is believed that all traces of their murder were obliterated by the militia on the huge funeral pyre."[31] No other source corroborates that claim. Yet it was not the UMWA but *The Rocky Mountain News* that coined the term "Ludlow Massacre" in an April 23 editorial.[32] In fact, the first official union estimate of the dead (as opposed to the estimate by strike leaders in the camp) was more conservative than that of many other sources.[33] However, miners throughout Colorado seized upon the deaths of women and children as motivation for the violent rampage that followed Ludlow. The deaths seemed particularly tragic to everybody who learned about them, but among miners they fed the highly organized, statewide campaign of revenge that has been largely overlooked in subsequent accounts of the massacre.

The federal troops that arrived in Colorado stopped the violence, but they could not stop the campaign against Rockefeller's reputation. As early as May 4, the managers of other Colorado coal firms protested: "We deplore the unjust attack upon Mr. Rockefeller. It is neither fair nor just to him nor to us to place the burden nor give him sole credit for the position we are maintaining."[34] Nevertheless, the United Mine Workers of America and its allies continued the attacks. The union and its representatives connected Rockefeller to Ludlow in countless public forums for months, even years, after the event. For example, the union sent two Ludlow survivors to the White House to confer with President Wilson, keeping the story in the news.[35] The survivors made many stops around the country before and after that meeting. These ladies (and their gender was no coincidence) were very effective spokespeople. *The New York Times*, covering the testimony of one of the survivors—Mrs. Mary Petrucci—before the Industrial Relations Commission, explained the reasons she was so effective as a union spokesperson. "She spoke good English and impressed the audience as a woman above her station," wrote the paper. "The fact that she told what happened with little display of emotion, as if anxious to get it over [with] as quickly as possible, made the telling all the more impressive."[36] Had Rockefeller attended the ceremony dedicating the Ludlow monument three years later, he would have heard exactly how the UMWA wanted Ludlow to be remembered. National UMWA president Frank Hayes composed and presented a poem for the occasion. It read, in part:

> But alas! There came a day.
> Greed demanded: "Stalk your prey,
> Fire the tents and shoot to slay!"
> Here on Ludlow Field.
> In the embers grey and red,
> Here we found them where they bled,
> Here we found them stark and dead,
> Here on Ludlow Field.[37]

Nobody in the audience needed to be reminded of whose greed Hayes meant.

The fate of union leader John Lawson also kept Ludlow in the public mind. Lawson, the most important local UMWA leader during the 1913–1914 strike, had been a union presence in the southern Colorado coal fields since 1906. While many believed members of the Colorado Militia should have been tried and convicted for murder after the Ludlow Massacre, a state-appointed grand jury in Trinidad instead indicted 124 striking miners—including Lawson—on various charges relating to violence before, during, and after the massacre. Many of the jury members had ties to coal companies, as did the judge who heard the case. Fred Farrar, Colorado's attorney general who called the grand jury, later served as corporate counsel to CF&I until 1952. Of all the indictments stemming from the strike, few reached the court. Only Lawson and three others were convicted, and the convictions were eventually overturned because of trial irregularities.[38] The case against Lawson stemmed from the killing of mine guard John Nimmo, which happened before the massacre occurred. The prosecution failed to produce any evidence that Lawson had pulled the trigger. In response to the conviction, the naturalist Gifford Pinchot declared, "If Lawson is guilty, not of actual murder, but for leading the striking miners, then Mr. Rockefeller, as the leader and employer of the murderous gunmen should be in the same cell as Lawson. If it is right for Lawson to go to jail for life, then I want to see John D. Rockefeller, Jr. go to jail for life."[39] Perhaps because of attacks such as this, Rockefeller tried to convince the governor to have the case against Lawson dropped during his first visit to Colorado. That effort failed, but the Colorado Supreme Court finally overturned his conviction in 1917.

Lawson's case made headlines around the country, making him a celebrity. Henry Ford even invited him to join his peace ship to Europe in its attempt to stop World War I. While he was in jail, southern Colorado miners elected Lawson president of their local.[40] Frank Walsh called him "one of the most splendid specimens of manhood in the United States."[41] The UMWA publicly called for Rockefeller's indictment on murder charges stemming from Ludlow as retaliation for Lawson's conviction.[42] Lawson became a problem for Rockefeller because many assumed that Rockefeller and CF&I had played a role in securing Lawson's conviction as a result of their political influence in southern Colorado. Rockefeller denied it, claiming "neither I nor any of my associates has had any connection whatsoever with this case."[43] The denial garnered yet more headlines.

After the United States entered World War I, the UMWA rhetorically linked the struggle for industrial democracy in Colorado and the deaths at Ludlow to the fight for democracy in Europe. "Let us keep their memory green," wrote the *Mine Workers Journal* shortly before the dedication of the Ludlow monument, "these humble soldiers, who gave up their lives in the great struggle for industrial freedom. We can pay them no higher tribute than that of giving our best service to the movement for which they died."[44] As World War I was raging at the time, the connection would have been obvious to every miner who read these words, especially as this was a common tactic employed by unions in the labor movement throughout the war. During President Wilson's 1919 Industrial Conference, when Rockefeller acknowledged, "Surely it is not consistent for us as Americans to demand democracy in government and practice autocracy in industry," he was in essence conceding that argument to the union.[45]

The UMWA received assistance in its protracted campaign against Rockefeller from a variety of sources, including the future head of the Committee of Public Information during World War I, George Creel, and former president Theodore Roosevelt.[46] However, perhaps the greatest source of outside help for the union was the muckraking author Upton Sinclair, who visited the site of the massacre shortly after it happened and concluded that John D. Rockefeller Jr. deserved to be publicly shamed for his role in the tragedy. Sinclair was later arrested for silently protesting outside Rockefeller's business offices while wearing a black armband. Sinclair addressed a series of meetings in Tarrytown, New York, near Rockefeller's mansion, at which hundreds of locals signed a petition asking President Wilson to nationalize the Colorado coal fields. Sinclair also came up with the idea of sending Ludlow survivors on tour across the country.[47]

To fight this campaign of blame, Rockefeller needed to appear both competent and compassionate. "The use of your own name in this affair has been most unfortunate," Lee wrote to Rockefeller during a Colorado fact-finding visit in August 1914. "It is important, in my judgment, that [the public's bad impression of you] be vigorously combated."[48] Unfortunately for Rockefeller, it took months for Lee and Rockefeller's other aides to implement the counterattack. At the beginning of the campaign in late 1914, Lee designed a series of pamphlets issued under the auspices of Colorado coal mine managers entitled "Facts Concerning the Struggle in Colorado for Industrial Freedom." Yet the errors and distortions in these bulletins (such as wildly inflating the salaries of UMWA strike leaders) led even UMWA president Jesse Welborn to try to convince Lee to issue corrections. "If the coal operators would lie about these things," wrote the UMWA in its own bulletin, "is it not reasonable to believe they would deceive you about other incidents connected with the strike?"[49] The pamphlets also hurt Lee's reputation when his authorship became public knowledge during hearings of the U.S. Industrial Relations Commission in January 1915.[50] "Ivy L. Lee—Paid Liar," wrote the poet Carl Sandburg in response. Upton Sinclair called him "Poison Ivy."[51]

The early mistakes by Lee's campaign and the difficulty of defending the deaths of women and children often forced Rockefeller and his associates to play defense. In July 1914, Lee wrote a bulletin headlined "NO 'MASSACRE' OF WOMEN AND CHILDREN IN COLORADO STRIKE" because "BOTH SIDES AGREE THAT NO WOMAN WAS STRUCK BY A BULLET FROM EITHER SIDE" (emphasis in original).[52] By focusing the operators' defense on the women and children, Lee was actually reinforcing the union's public relations strategy. In addition, with pictures of the pit in which the women and children suffocated splashed on front pages throughout the country, Lee's premise was simply beyond belief. Similarly, when confronted directly about Ludlow before the U.S. Industrial Relations Commission, Rockefeller protested, "The emphasis had always been put upon the women and children killed in the ground. . . . [T]hey were smothered, and not struck [by bullets]."[53] Even though Rockefeller was correct, such nitpicking seemed callous.

ORIGINS OF THE ROCKEFELLER PLAN

The Colorado Fuel and Iron employee representation plan was a direct result of the bad publicity the company received in the wake of the massacre. "The plan first took form in Mr. Rockefeller's mind," explained George West in the report on the tragedy prepared for the U.S. Industrial Relations Commission, "when, after the Ludlow massacre, aroused public opinion frightened him into a realization that something must be done."[54] Rockefeller's first documented move toward employee representation came when he asked Mackenzie King for advice in an August 1914 letter. To restore "a permanent condition of peace," Rockefeller wanted King to suggest an outline of "some organization in the mining camps which will assure to the employes the opportunity for collective bargaining, for easy and constant conferences with reference to any matters or differences or grievances which may come up, and any other advantages which may be derived from membership in the union." This statement shows not only that Rockefeller considered "collective bargaining" to include bargaining without an independent trade union but also that he intended the ERP to be a union substitute. King, busy with Canadian government matters, merely suggested a "board on which both employers and employees are represented, and before which, at stated intervals, questions affecting conditions of employment can be discussed and grievances examined."[55] Later, when Rockefeller first met King in person, he was so impressed that he immediately offered him a job with the Rockefeller Foundation.[56] Only when King joined the foundation on October 1, 1914, did work on an exact plan of employee representation begin in earnest. Yet even before that date, King worked hard to convince Rockefeller of the need for drastic changes in CF&I's industrial relations policy. King believed the plan he was incubating could both forestall union organizing drives and eliminate the need for future strikes—two outcomes Rockefeller very much wanted.

While an ERP had been discussed within CF&I management circles for months, the idea had yet to be broached in public. In private, some CF&I executives resisted implementing an ERP while the strike continued, but that resistance disappeared when the federal government threatened to become directly involved in the continuing dispute between management and the UMWA. In July 1914 the UMWA proposed settling the strike by waiving union recognition for three years but establishing formal machinery to ensure that miners' grievances would be addressed.[57] On September 5, 1914, President Wilson made public his similar plan for a truce. The UMWA accepted the truce immediately, even though it fell short of the union recognition it wanted.[58] Company president Jesse Welborn, in a public letter to the White House, politely refused to accept the truce, citing threats of violence from the UMWA and alleged inequities in the proposed settlement.[59] On December 1, Wilson issued a statement that blamed Colorado coal operators for the failure to settle the strike.

The statement also announced the appointment of a grievance commission (popularly known as the Low Commission), headed by National Civic Federation president and former New York City mayor Seth Low, to investigate the conflict and mediate future disputes around the country. In the statement, Wilson expressed his hope that "both parties may see it to be not merely to their own best interest but also a duty which they owe to the communities they serve and to the Nation itself to make use of this instrumentality of peace and render strife of the kind which has threatened the order and prosperity of the great State of Colorado a thing of the past."[60] When it appeared that failing to offer an alternative to unionization might help the union gain public support for the federal government to intervene on its behalf, management took up King's proposal for employee representation in earnest. As his aide Starr Murphy had written to Rockefeller back on September 16, "It seems clear to me that public opinion will demand either the acceptance of the President's proposition, or some constructive suggestion from the operators. A mere refusal to do anything would be disastrous."[61] That "constructive suggestion" became the Rockefeller Plan.

The UMWA formally ended its strike on December 10, 1914, citing Wilson's statement when he created the Low Commission as cover for its retreat. "In view of this urgent request, coming as it does from the Chief executive of the Nation," read the UMWA's statement, "we deem it the part of wisdom to accept his suggestion and to terminate the strike." Nevertheless, the union leadership could not resist one last shot at John D. Rockefeller Jr.:

> All lovers of liberty and believers in fair play between man and man must admire the heroic struggle of the Colorado miners against the great wealth and influence of Rockefeller and his associates. We believe that our people have not died in vain, and that the battle they have waged against such tremendous odds has aroused the conscience of a Nation, and that out of the martyrdom of our people will come the dawn of a better day for the suffering miners and their families.[62]

The UMWA organized nine separate locals in the southern Colorado coal fields between the end of the strike and Labor Day 1915.[63] If management did not do something drastic, CF&I faced the immediate prospect of another dispute. Rockefeller's first action in this regard was to ask for Lamont Bowers's resignation shortly after the strike formally ended in December 1914. Bowers, a fierce opponent of trade unions, had been the public face of the company since the strike started in 1913 and had been Rockefeller's eyes and ears in Colorado during the conflict. Retaining the title of president, Jesse Welborn also took over Bowers's functions as chairman of the board. Unlike Bowers, Welborn expressed a willingness to support employee representation, and he had been less publicly identified than Bowers with the strike and the resulting tragedy at Ludlow. This change at the top, therefore, was a way to show both employees and the public that new policies were forthcoming.

At the same time, CF&I instituted a stopgap employee representation plan intended to forestall interference in its industrial relations by the Low Commission. On December 16, 1914, Welborn appointed David

Griffiths—a well-respected former CF&I miner, superintendent, and state mine inspector—as the first president's industrial representative, to act as an intermediary between labor and management. On January 5, 1915, management posted notices in every coal camp asking miners to select one representative for every 250 employees. That notice read, in part:

> The employees of the Colorado Fuel and Iron Company employed at ____ are hereby entitled to assemble in Mass Meeting on the ____ of ____ at the hour of ____ at ____ for the purpose of selecting by ballot one or more of their number to represent them at a Joint Meeting of themselves and representatives of the Company, to be held in Denver, for the purpose of discussing matters of mutual concern and of considering means of more effective co-operation in maintaining fair and friendly relations.[64]

The elections were completed by January 19.[65] The plan was entirely company-initiated. While this distinction would not be important until the passage of the National Labor Relations Act in 1935, it explains much of the plan's subsequent history. Welborn seemed happy with this system as it had evolved to that point, but King thought otherwise. "I think Mr. Welborn's plan is a mere beginning, which may come to be viewed as little more than a formality," he wrote.[66] Nevertheless, Griffiths would begin his efforts to resolve differences between labor and management while King worked on the final version of the industrial plan that would be formally introduced in October.

Rockefeller, King, and their associates in Colorado obviously intended that the Rockefeller Plan would counter the extraordinarily bad publicity CF&I had received ever since the coal strike began in 1913. Rockefeller told members of the Colorado Justice League that "he had been charged with responsibility for crimes . . . which he had not even heard of at the time of their occurrence, and he seemed truly grieved that people should think that of him." This explains why he aimed "to do everything in [his] power to prevent [the] recurrence" of the Ludlow Massacre.[67] To stave off the possibility of future Ludlows and restore Rockefeller's reputation, the illusion of change simply would not do. The company's labor reforms had to be effective. Even if somehow the public were to stop paying attention to events in southern Colorado, the United Mine Workers of America would continue to monitor them. To keep the union out and to avoid outside pressure on their management decisions, CF&I leaders had to offer real change, even if in a closely controlled manner.

JOHN D. ROCKEFELLER JR.'S FIRST COLORADO TOUR

One of the most important events in Rockefeller's rehabilitation campaign was his testimony before the U.S. Industrial Relations Commission. At public hearings in New York City in January 1915 and in Washington, D.C., that May, he faced blistering questions from commission chair Frank Walsh. As John Fitch described it, "Young Mr. Rockefeller, if we are to accept his statements upon the witness stand, knew very little of what was going on in Colorado. Almost without exception he protested his ignorance concerning pertinent facts, of which every newspaper-reading American has heard something."[68] Even though many people criticized Rockefeller for the extent of his ignorance on matters concerning CF&I, many others were generally sympathetic after the tough questioning ended. "I should hope," Rockefeller replied in response to one of the easier questions posed to him, "that I could never reach the point where I would not be constantly progressing to something higher, better—both with reference to my own acts and . . . to the general situation in the company. My hope is that I am progressing. It is my desire to." These remarks brought forth a round of applause in the hearing room.[69]

One of Rockefeller's sympathizers was Mother Jones, the legendary "miners' angel," recently jailed twice in southern Colorado jails for supporting the strikers. At the close of his second grueling day of testimony, she told Rockefeller, "I liked your testimony very much." He replied, "Thank you. I think I should like to try to work with you in some of your efforts." She responded, "Well come out to Colorado and we'll show you some things you ought to see."[70] Later, she told the assembled press, "I just told Mr. Rockefeller one thing. We have been misrepresenting him terribly, and I as much as anybody else."[71] Jones made the same positive assessment of Rockefeller's character in private. "I believe the human is very deeply planted in his breast," she wrote an associate two months later. "[W]hen he understands the conditions under which his people suffered here [in Colorado], he will do everything to remedy the wrongs."[72] If as committed a supporter of labor as Mother Jones was swayed by Rockefeller, there was probably something serious behind his words.

At this juncture, Rockefeller made the shrewdest move he could have made in the face of all the criticism: he admitted his ignorance concerning labor issues and explained to the world what he was going to do to remedy the situation. Immediately after his meeting with Jones he told reporters, "Gentlemen, I know it is my duty as a director to know more about actual conditions in the mines. I told Mother Jones that, of course, there should be free speech, free assembly, and independent, not company-owned schools, stores and churches in the mine field. I am going to Colorado as soon as I can to learn for myself."[73] Before Jones had suggested the trip, the idea of Rockefeller going to Colorado had been circulating among his advisers since shortly after the massacre, as revealed in the way he framed his announcement. Rockefeller explicitly linked going to Colorado with the recently ended strike to help put the strike (and, by extension, the massacre) behind him. The trouble had occurred, he implied, because he did not understand what was happening on the ground in Colorado. Now he would know. He introduced the Rockefeller Plan during that trip to ensure that management would always understand the conditions its employees faced. That way, strikes that arose as a result of failures in communication would presumably never happen again.

In anticipation of his trip, in September 1915 Rockefeller, King, and Starr Murphy met with John P. White, president of the United Mine Workers of America, and William Green, secretary-treasurer of the American Federation of Labor. Rockefeller told the labor leaders about the still-incubating Rockefeller Plan, and they both approved it heartily. "I took it for granted that what they were driving at was an agreement with the United Mine Workers of America," the younger Rockefeller wrote to his father. "We discussed the situation very fully, however, and these men made it clear that they did not believe such an agreement was practicable or possible at this time. In fact, Mr. White said to Mr. King before leaving that if such an agreement could be made by our Company with its own employees, they would not ask for any agreement with the union now."[74] If White and Green had seen the plan as an impediment to unionization, they never would have offered their approval. They also encouraged Rockefeller to visit Colorado and see the miners firsthand.[75]

Despite appearances to the contrary, Rockefeller had not really forsaken anti-unionism.[76] As he explained to his father:

[O]nce the employees are thoroughly convinced of the sincerity of the purpose of the Company to insure in definite, tangible form to every man redress for any real grievances, a proper regard for his right to appeal and the establishment of a permanent channel of communication between the duly appointed representatives of our employees and the officers, the probabilities of being willing to pay their dues to the United Mine Workers for not better, if as good, protection of their rights than this agreement will assure them, would seem to us reasonably small. Even if, however, the men should be very generally brought into the union, we will be assured of quiet operation during the life of the agreement, and at its expiration, if the vast majority of the men who have been giving satisfaction as our employees for some years past should prove to be members of the union, it would be a question then for us to consider whether we should not recognize the union, so long as men who are non-union were not discriminated against.[77]

Rockefeller believed his and King's union could beat the United Mine Workers of America at its own game, offering the company's employees better representation at no cost to them. This was an anti-union strategy, but it was different from the anti-union strategies of most industrialists who did not really care whether their employees' concerns were addressed. This fact helps explain why Rockefeller received such good press when he first visited Colorado in late September–early October 1915.

"All the world knows what John D. Rockefeller, Jr., was doing last week in Colorado," explained *The New York Times* near the end of his trip.[78] Pictures of Rockefeller dancing with miners' wives helped humanize a man whose recent press clippings had been even worse than his father's at their worst. "My wife's 2000 miles away," he told a crowd assembled at that stop, "and I feel like having some fun."[79] The dancing lasted until shortly before midnight.[80] While some praised Rockefeller for his bravery because of his willingness to travel among the miners without a bodyguard, his real bravery lay elsewhere. Rockefeller was an extremely shy man, who later in life became something of a social recluse. He had little experience mixing with strangers, let alone the working-class people he met on the trip. His performance was not only masterful from a public relations standpoint but must also have been incredibly difficult for someone so shy. Without Mackenzie King, who traveled with him throughout the trip and took on the job of making Rockefeller interact with the public, the tour would have been much less successful.[81]

1.2. *Mackenzie King (center) and John D. Rockefeller Jr. (right) don miner's garb during Rockefeller's first tour of Colorado in 1915. The miner on the left is employee representative Archie Dennison. Courtesy, Bessemer Historical Society.*

Rockefeller used the tour both to distance himself from the strike and to set the stage for the cooperative venture that was to come. At each mine camp he visited, Rockefeller held a conference with the local representatives elected under the tentative version of the plan implemented the previous winter. "Look here, Mr. Morelli," he told one miner. "I want to have a heart to heart talk with you. Just forget that I am a wealthy man, and talk frankly to me as you would to any one of your brother miners."[82] This was the principle behind the Rockefeller Plan writ small: miners and management working things out informally. By Rockefeller's account, his talks with miners went extremely well. As he recalled in a speech in Cleveland the following year:

> I found no single instance where the men, as I met them, were other than friendly, frank, and perfectly willing to discuss with me, as I was glad to discuss with them, any matters they chose to bring up. It frequently occurred that there was justice in the points which they brought up, and their requests were acted upon favorably by officers. Also frequently situations were presented in which it was impossible for the company to meet the views of the employes. But never was the subject dismissed until, if unable to make this situation clear, the highest officials in the company were called in to explain to employes, with the utmost fullness of detail the reasons why the things suggested were impossible.[83]

Rockefeller later claimed he did not hear a single negative comment from a worker during the trip.[84]

"Rockefeller wins over miners who forget tragedy at Ludlow," read one headline in *The Denver Post*.[85] "Rockefeller turns hate of miners to love," read another from *The Chicago Tribune*.[86] Such sentiments, however, did not convey the context of the entire visit. For one thing, Rockefeller did more than talk during the trip. For example, Rockefeller told a miner from Sopris to "[g]o ahead and build the bandstand and a park and send me the bill."[87] At one camp, Rockefeller distributed personal checks of ten or fifteen dollars as prizes to miners' families whose houses had the best gardens.[88] This begs the question of whether the miners who loved Rockefeller did so for himself or for his money. Furthermore, in an error of judgment CF&I management often reprised when dealing with workers throughout the life of the plan, workers did not always express their true feelings to Rockefeller. Did Rockefeller consider this exchange between himself and African American miner Willis Hood (as recounted in *The New York American*) to be positive or negative?

[Hood]: "And this is what I wants to say to you, Suh. I have been a working for you for twenty-three years. Now, how about a pension? When are we going to have pensions so I can quit work?"

"Well, I don't know," said Rockefeller, "but when we get the system ready, we will both apply."

"What you?"

"Why yes. I don't get a pension."

"But you don't do no laborious labor," cried Hood, "and you don't need no pension. Besides we are looking to you as the source from which the pension is to come."[89]

The resentment here is palpable just from reading the transcript.

Rockefeller met other workers who thought like Hood did, whether he knew it or not. Sixty years later Dan De Santis, a mule driver at the same mine, remembered his conversation with Rockefeller as he drove the industrialist to the coal face. Rockefeller asked, "'Do you drive this team?' I said, 'yeah.' 'You do that all day long?' I says, 'That's all I got to do: I got to make a living.' And he says, 'Good boy.' But they wasn't paying me nothing, see."[90] The tour may have brought good publicity for Rockefeller, but it did not necessarily make the workers happy. Some workers wanted more than management had offered them to that point. A visit from Rockefeller could only go so far in improving relations, no matter how much the miners liked him personally.

The highlight of the tour was Rockefeller's formal introduction of the Rockefeller Plan in a speech outside the Minnequa Works in Pueblo on October 2 and a subsequent meeting with the miners who served as the first representatives. Rockefeller called it a "red-letter day" in his life. The press, stressing Rockefeller's apparent sincerity, applauded his goals in glowing terms. *The Denver Post* introduced its story on the Pueblo speech this way:

> Granting practically every point which any labor union ever asked, with the one exception of recognition of the union, which he says is not an impossible thing in the future, and voluntarily offering certain betterments in advance of union demands . . . John D. Rockefeller, Jr., yesterday, at Pueblo, gave out to a meeting of miners representing his employes the text of the "Rockefeller industrial plan," by which he and his associates hope to bring about permanent industrial peace in the mining camps of the Colorado Fuel & Iron company.[91]

1.3. *John D. Rockefeller Jr. (foreground center) surrounded by the first group of employee representatives during his visit to Pueblo to announce the creation of the Rockefeller Plan, October 1915. Courtesy, Bessemer Historical Society.*

The *Post* reprinted the entire text of the plan. *The Fort Collins Weekly Courier* editorialized, "If John D. Rockefeller, Jr.'s, visit and inspection of conditions in southern Colorado results in an agreement between mine workers and the operators whereby industrial peace in the coal mining region will be maintained in the state, as now seems probable, there will be great rejoicing all along the line. . . . [H]e will be hailed as one of the state's greatest benefactors."[92] Sadly, however, it was not to be.

The *Courier* likely picked up the possibility of the Rockefeller Plan bringing lasting industrial peace from Rockefeller himself. Rockefeller made his aspirations for the plan abundantly clear throughout the tour. For example, in a speech to the Denver Chamber of Commerce on October 8, Rockefeller declared:

> [I]t is the earnest hope of all who are associated in the plan that it
> may point the way to a closer co-operation between the employes and
> the other parties [with] interest in this company, that it may establish

relations of friendship and of mutual confidence, that it may so benefit the workers, the officers and the stockholders of the company, that there may never come a day when there will be repeated the industrial disorders which have occurred in the past in this company and other companies in this State.

Rockefeller's aspirations did not stop at the Colorado state line. "[I]f in any smallest way my coming to Colorado may prove to have been [of] service to you in approaching the solution of this world problem of industrial relations," he told the audience, "I shall feel a sense of satisfaction and gratitude beyond expression."[93] This was Rockefeller's motivation in a nutshell: he wanted to solve the labor problem not only because of his own interests in Colorado Fuel and Iron (which, of course, were miniscule in relation to his family's overall fortune) but also to repair his damaged reputation. In this way, Rockefeller could turn a strike in Colorado that made him appear autocratic into a launching pad for bringing industrial democracy to the masses and thereby carve out a niche in life independent from his father.

Company managers recognized that they had to submit the Rockefeller Plan to their employees for a vote if workers were ever to take the idea of industrial democracy seriously. Yet there were no dissenting voices against the plan at that first meeting in Pueblo. Urged to talk about the plan among themselves, the miners approved it unanimously without discussion.[94] Their confusion about the idea of deliberating and voting in this situation is understandable, as the concept of workers voting on anything having to do with their jobs was fairly novel in 1915. Nevertheless, the miners had taken cues from management in this instance. Rockefeller had pitched the creation of the plan as a give-and-take process during his Pueblo speech, yet management gave the miners no means by which to make suggestions, let alone to construct an entirely new plan from scratch. The vote was up or down, both in Pueblo among the representatives and in the camps where the rank-and-file miners voted.

Management wanted people to think that the employee representatives had explained the plan to their constituents back in the camps. In fact, members of management did the explaining. After that, the miners voted "yes" or "no." Recognizing that many of the miners were illiterate (at least in English), camp officials gave them colored ballots to use in voting on the plan. White meant yes; red meant no. The election process was designed so management could take great pains to demonstrate that workers were not pressured to vote "yes." The company had the votes tabulated by October 20, 1914. The results strongly suggest that no direct pressure had occurred. While a large majority of miners (2,404) voted in favor of the plan, 442 miners voted "no."[95] This was an impressive majority, but perhaps more important, only 65 percent of CF&I miners bothered to vote at all—2,846 miners out of a possible 4,411. Indeed, according to Ben Selekman and Mary Van Kleeck, who computed these figures from company records, "very active efforts [had] to be made to get even this number of men to vote, and it is unfortunate [for] the success of the plan in stimulating the initiative of the wage-earners that these efforts [were] made by officials rather than the miners themselves."[96] According to Rockefeller, the results indicated that "the men in the mines believe that we are sincerely in earnest in our desire to better conditions of the mines and to do all that is possible to promote the conditions of employment, and the living and working conditions of the men."[97] In private, Rockefeller indicated in a memo covering his entire trip to Colorado that David Griffiths told him "it was almost impossible to get the men to take an interest in the meeting."[98] Things improved little during the life of the plan.

Chapter Two

STUDENT AND TEACHER

> As you know all my life I have sought to stand between labor and capital, trying to sympathize with and understand the point of view of each, and seeking to modify the extreme attitude of each and bring them into cooperation.
>
> —JOHN D. ROCKEFELLER JR. TO HARRY EMERSON FOSDICK, DECEMBER 19, 1927[1]

When John D. Rockefeller Jr. introduced the employee representation plan (ERP) that would bear his name before an audience of miners in Pueblo, he explained that four parties are involved in every corporation: the stockholders, directors, officers, and employees. He then compared Colorado Fuel and Iron (CF&I) to a table:

> This little table [*exhibiting a square table with four legs*] illustrates my conception of a corporation.... First, you see that it would not be complete unless it had all four sides. Each side is necessary: each side has its own part to play....
>
> Then, secondly, I call your attention to the fact that these four sides are all perfectly joined together.... Likewise, if the parties interested in a corporation are not perfectly joined together, harmoniously working together, you have a discordant and unsuccessful corporation.
>
> Again, you will notice that this table is square. And every corporation to be successful must be on the square—absolutely a square deal for every one of the four parties, and for every man in each of the four parties.

I call your attention to one more thing—the table is level. Each part supported by its leg is holding up its own side, hence you have a level table. So equal responsibility rests on each of the four parties united in a corporation.

Rockefeller extended this metaphor by piling coins on the table to represent the earnings a square, level corporation can accumulate.[2]

Speaking before the Denver Chamber of Commerce a few days later, Rockefeller made a special effort to contrast CF&I's new ERP with the operation of "some organizations of labor" (meaning independent trade unions), which he said arrayed labor against capital, brought war to the workplace, and operated so that "only those who elect to join an organization are eligible [for] the benefits that come with it."[3] He made no such contrast in his remarks to CF&I employees.

This was a new kind of Rockefeller—a better friend to labor than even unions were. However, the Rockefeller Plan was no charity scheme. John D. Rockefeller Jr. thought improving the terms and conditions of labor at CF&I by itself would lead to higher profits. At the time, labor and other capitalists derided this idea, but modern economic research suggests Rockefeller may have been right. Allowing employees to express their voice in the workplace can be good business under some circumstances. In the economic sense, "voice" refers to employees' willingness to tell management their problems so the problems can be fixed rather than have the employees quit and find better jobs. Economists have developed this concept to demonstrate the potential positive effects of independent trade unions. David Fairris, who examined a number of ERPs from the 1920s, found that they had similar effects—decreased turnover, increased productivity, and fewer worker injuries.[4] Richard B. Freeman and Edward P. Lazear found similar positive effects from modern incarnations of ERPs.[5] While CF&I profited greatly under the Rockefeller Plan during good times, those gains could not get the company through the dire straits it faced in the mid-1920s and 1930s. In a seminal 1980 article in the *Quarterly Journal of Economics*, Freeman explained how arrangements such as the Rockefeller Plan improved morale but did not lead to a recognition of mutual interest among a firm's employees. "The dilemma is that if management gives up power, it creates seeds of genuine unions," he suggested. "[I]f it does not, employee representation plans face severe difficulties."[6] CF&I management made many concessions to employees, but it never ceded significant power without resistance.

Despite this failing, the Ludlow Massacre nonetheless changed John D. Rockefeller Jr.'s thinking about industrial relations, as is obvious from his many articles and speeches on the subject. "Ludlow was a rite of passage for Father," Rockefeller's son David explained in his 2002 memoir. "Although not a businessman by talent or inclination, he had demonstrated his skill and courage." More important for this study, David Rockefeller believed that "Father's objective [in introducing the Rockefeller Plan at CF&I] was to improve labor relations in the United States by addressing the grievances of labor and persuading businessmen to recognize their broader responsibilities to their workers. For that reason his involvement with labor issues did not end with Ludlow but remained a central issue for the rest of his life."[7] Colorado Fuel and Iron would serve as Rockefeller's most important laboratory for the principles Mackenzie King taught him in the years following the tragedy.

As explained in Chapter 1, the Rockefeller Foundation hired King as an adviser shortly after the massacre to help make Rockefeller's ideas about developing a non-union labor organization a reality. "That King was a liberalizing influence [on Rockefeller] is obvious," wrote Raymond Fosdick in notes for his authorized biography of Rockefeller, written during the subject's lifetime.[8] Rockefeller's relationship with King was close and surprisingly personal for such a proper and formal man.[9] King tutored Rockefeller on the principles of industrial relations generally and the mechanics of employee representation specifically. King wrote the text of the "Rockefeller Plan"; Rockefeller supplied the power to make it happen and the vision to see it into fruition. As he became more familiar with them, Rockefeller became the chief spokesperson for King's ideas on labor. In 1906, the legendary Mark Twain (who knew and professed to like Rockefeller) wrote, "Young John has never studied a doctrine for himself; he has never examined a doctrine for any purpose but to make it fit the notions which he got second hand from his teachers."[10] That observation may have been true when Twain wrote it, but it was not true with respect to employee representation. Rockefeller did not slavishly copy all of his ideas on this subject from King; indeed, King later left the Rockefeller Foundation and returned to Canadian politics. Therefore, on the pivotal question of whether to recognize independent unions, Rockefeller developed ideas totally on his own.

In 1915, few American companies made even rhetorical concessions to the notion of worker empowerment. As Ben M. Selekman and Mary Van Kleeck (no friends of the Rockefeller Plan) explained in 1924, "[I]n advocating representation for wage-earners, Mr. Rockefeller is definitely opposing the traditional view that management alone, with its knowledge of business conditions, should establish the terms of employment without any organized means of consulting the workers."[11] Of course, John D. Rockefeller Jr.'s magnanimity was motivated in part by self-interest, something Rockefeller himself admitted.[12] Nevertheless, Colorado Fuel and Iron lost far more money than it made during the years the Rockefellers controlled it. While this performance obviously cannot be blamed exclusively on the plan and the costly welfare capitalism that accompanied it, Rockefeller's comparative benevolence carried significant financial risks from an industrial relations standpoint. He conceived of and pitched the ERP to workers and the general public as a means to achieve permanent industrial peace. Therefore, this is the standard by which the Rockefeller Plan deserves to be judged. By that standard, it fell far short of its goal because too many CF&I employees wanted more than the Rockefeller Plan offered them.

JOHN D. ROCKEFELLER JR. TAKES ON THE LABOR QUESTION

John D. Rockefeller Jr. was a rather ordinary-looking man of medium height, somewhat stocky, and with no distinguishing features. As the son of a world-famous father, he had been in the public eye since his early childhood. Despite that fact, few people recognized him. When the governor of Georgia stayed in the hotel in which Rockefeller was staying during the industrialist's 1915 Colorado tour, he had been briefed about Rockefeller's appearance both by newspaper coverage of his travels and by the staff at Denver's Brown Palace. "Is Mr. Rockefeller in the hotel?" the governor asked in his quest for an introduction. "Why, he just walked past you," came the response.[13] Such incidents might have been at least in part a result of Rockefeller's anti-social streak and self-effacing personality. Indeed, as many writers have pointed out, Rockefeller's willingness to be called "Junior" his entire life says a great deal about his self-image. Rockefeller was not a great public speaker, so much of the positive press he received during his 1915 Colorado tour undoubtedly resulted from his humility rather than his charm.

Rockefeller's lack of gravitas might explain why Fosdick's biography of his longtime friend and employer remains the only full treatment ever published about the industrialist's life. The biography paints an overwhelmingly positive portrait "of a man who through adversity rose to eminence and even greatness."[14] According to Fosdick, the "adversity" Rockefeller faced was wealth. This twist on an old model for a biography allows Fosdick to tell his tale of a born millionaire in the same manner that Andrew Carnegie and Benjamin Franklin told their life stories to earlier generations. Interestingly, David Rockefeller offers a much more revealing portrait of John D. Rockefeller Jr. in his memoir. "Father was a complicated person," he wrote:

> Grandfather was a self-made man who created a great fortune starting with nothing, an accomplishment Father would have no opportunity to emulate. Even after he had built a solid record of achievement, he was plagued with feelings of inadequacy. He once described his brief involvement in the business world as one of many vice presidents at Standard Oil as "a race with my own conscience," and in a sense Father was racing all his life to be worthy of his name and inheritance.[15]

While most of us can only hope to be faced with the kind of "adversity" Rockefeller had to confront, his extraordinary wealth gave him the freedom to think about the kinds of problems most businesspeople never bothered to consider.

One of the biggest problems of the era was the so-called labor question, how to get workers to accept the difficult circumstances created by industrialization. While Rockefeller strongly hinted at his aspirations to solve this problem during the first Colorado tour, King was even more explicit. In a letter to Rockefeller's wife, Abby, he wrote, "I cannot but feel that this visit is epoch-making in his [Rockefeller's] own life, as it will also prove epoch-making in the industrial history of this continent. From now on he will be able to devote his time to advancing the vast projects . . . [related] to human well-being."[16] For a man who could not prove himself by pulling himself up by his bootstraps, finding solutions to intractable problems such as this one became a way to make a unique contribution to the world. Besides advocating employee representation as a solution to the labor problems caused by industrialization, Rockefeller funded the creation of Industrial Relations Counselors, Inc., and the Industrial Relations Section at Princeton University to address the same problem. In his history of the entire field, Bruce Kaufman traces one of two main branches of industrial relations—the personnel management school—to Rockefeller's influence.[17]

The great irony regarding these efforts to solve the labor question is that very few people had played a bigger role in creating the problem than Rockefeller's father, who literally fueled America's industrial revolution. John D. Rockefeller Sr. became rich beyond most people's wildest dreams by building the Standard Oil Company up from a regional to an international enterprise during the late nineteenth century. The primary ethical controversy surrounding the Rockefellers' money concerned Senior's actions toward other businesses, not his attitudes toward labor. Still, Rockefeller Senior was fiercely anti-union, as was his son in his early years. During a 1903 organizing campaign, the younger Rockefeller wrote CF&I's president, "We are prepared to stand by in this fight and see the thing out, not yielding an inch. Recognition of any kind of either the labor leaders or [the] union, much more a conference such as they request, would be a sign of evident weakness on our part."[18] This statement, as well as the younger Rockefeller's infamous congressional testimony about the open shop being a "great principle," reflected a simple-minded adherence to the rhetoric of individuality that countless other businesspeople expressed during that time.[19]

While he never became a supporter of independent trade unions, Rockefeller's experience after Ludlow led him to take public positions on labor issues that virtually no industrialist of his stature would even have considered. Rockefeller went from believing the open shop was worth "killing all your men" for to believing labor and capital shared an identical interest in seeing that a business succeed. For their part, most of Rockefeller's capitalist contemporaries still clung to the notion that workers deserved nothing but the opportunity to quit if they found the terms and conditions of their employment unsatisfactory. American businesspeople bore no legal obligation to recognize trade unions; therefore, few did so.

To demonstrate the contrast between the post-Ludlow Rockefeller and his contemporaries, it is useful to compare Rockefeller's attitude toward labor to that of Charles M. Schwab, the chair of Bethlehem Steel, whose firm also hired Mackenzie King to design its ERP. In 1920, Schwab told the New York Chamber of Commerce, "Labor should have its fair share of the results of industry. Labor should be recognized as entitled to consult with management in the mutual interest. Labor cannot be driven, and business cannot be successful unless the men employed in it are enthusiastic and loyal. That loyalty cannot be obtained with a big stick; it must be based upon fair dealing and sympathy." While his ideas were superficially similar to Rockefeller's, Schwab made no pretensions of democracy as Rockefeller did. "I will not permit myself to be in a position to have labor dictate to management," he explained in 1919, after the firm's ERP had launched.[20] No wonder King distanced himself from Bethlehem Steel's ERP. As Howard Gitelman has explained, management there "had no intention of seriously consulting or listening to its employees."[21]

Rockefeller's earlier unrelenting hostility toward trade unions was also a product of his complete lack of interest in conditions on the ground in Colorado. "Mr. Rockefeller is a long-distance director," explained the report of the congressional committee that investigated Ludlow, "not having attended a meeting of the stockholders or directors of the company in 10 years."[22] This is the reason that Rockefeller had to wire Lamont Bowers following the massacre to ask how many other coal companies were involved in the strike and what percentage of the market CF&I controlled.[23] In 1915, Rockefeller admitted before the U.S. Industrial Relations Commission, "I have never had the personal handling of labor questions. I have had such matters before me, but beyond making what I have tried to have be a very complete statement of my position, I do not feel sufficiently qualified to discuss intelligently and usefully the details relating to the matter [of whether workers should join unions]."[24] Material in both the Rockefeller Archives and the Colorado Fuel and Iron Archives suggests that Rockefeller was determined to never again have to plead ignorance about labor relations in his mines.

After returning from Colorado in 1915, Rockefeller took a personal interest in exactly how management implemented his and King's creation. Henry Hall of *The New York World* witnessed this enthusiasm while interviewing Rockefeller a month after he returned to New York. Rockefeller "had the entire plan in all its details right at his fingers' ends," Hall reported. "[T]he manner in which he presented it showed not only that he knew every phrase of it but that he believed it."[25] Rockefeller showed this same interest in private correspondence as well. At around the same time as the *World* interview, his aide Charles Heydt (who had accompanied Rockefeller on the first tour of Colorado) wrote to CF&I executive Elmer Weitzel at Rockefeller's request:

> Mr. Rockefeller is wondering if the checkweighman [whose job was to weigh the coal individual miners dug] employed at Berwind at the time of his visit is still in that position? If not, when did he leave?
> Have the wire fences been erected between houses at Berwind?
> Have the old red houses at Berwind been replaced or repaired?
> Have fences been built about the houses at Tabasco? There were no fences at the time of Mr. Rockefeller's visit.
> At Trinidad a suggestion was made that the retail grocery store might be sold. Has this been done?[26]

From then on (at least until 1919), the director of welfare work at CF&I compiled and sent Rockefeller a summary of all the grievances men throughout the company had lodged (whether formally or not) with their elected representatives.[27] Rockefeller used these reports as a basis for inquiries and suggestions. "On November 19th," wrote Rockefeller in one such instance, "the summary states that at Fremont, representative Ossolo had not been at work since his accident. I shall be interested to know what the accident was, how serious, and whether Ossolo is well again. When [President's Industrial Representative David] Griffiths is at Fremont again, perhaps he will mention to Ossolo my regret at his accident and my hope that he has entirely recovered by this time."[28] Other inquiries offered suggestions designed to make the ERP more effective. "I am wondering whether the hearing at Walsen on the complaint of Rosetti was conducted in such a way as Rosetti would feel perfectly free to talk," he wrote in July 1916. "It seems to me that [if] there were such a charge made Mr. Griffiths should interview the person privately, and give him an opportunity to present whatever evidence he has in the absence of the accused foreman."[29] Rockefeller oversaw—some might say micromanaged—the implementation of the plan because he wanted it to work. After all, it had his name on it.

Besides his sincerity, Rockefeller's concern for workers' anonymity demonstrated his understanding that CF&I employees had the power to make the plan a success or a failure by the way they responded to it. To ensure a positive response, management had to meet its obligations. As he explained very simply in a 1915 letter to his father: "[The agreement] must be lived up to or else be a farce."[30] Therefore, Rockefeller pressured his managers to adhere to his ideals. "There may be a tendency to hold district meetings or committee meetings less often than is planned or is desirable," he wrote CF&I vice president Arthur H. Lichty in 1921. "It may often be easier to have matters taken up with an officer of the Company than through the committees appointed for that purpose.... Many other matters of this kind there are which some one will need to see to constantly in order to keep the machine working completely and most effectively."[31]

Rockefeller witnessed evidence of success during his second trip to the CF&I properties in 1918. "The men seem to have reached a perfect agreement with their employers under this industrial plan," he declared while touring the Minnequa Works that year. "I have never seen such harmony between thousands of working men and their employers.... I talked with these men when there were no 'bosses' around to make the men afraid to speak out."[32] Of course, Rockefeller himself was a boss—the boss of bosses. Was the industrialist really too naive to realize this? The answer is probably yes, as even the new pro-labor Rockefeller remained an unreconstructed elitist. As Mother Jones explained to a reporter in an otherwise friendly assessment of Rockefeller's testimony before the U.S. Industrial Relations Commission in January 1915:

> Did ye note ... what he said when Chairman [Frank] Walsh asked him if he thought it was fair that the workmen, who suffered and were maimed in the mines, should not get more of the returns than they did in comparison with his father, who never went near the mines? He answered promptly that capital deserved a good percentage. He showed plainly that it was in his bones that capital was divinely appointed to dole out a little living to labor.[33]

Rockefeller maintained this attitude long into his enlightened stage on labor issues, writing of the "common man [sic]" in 1934, "They have very little money but are delightful."[34]

"KING IS OUR AUTHORITY NOW"

"Seldom have I been so impressed by a man at first appearance," Rockefeller later said of William Lyon Mackenzie King.[35] This should come as no surprise because the two had much in common. They were approximately the same age. Like Rockefeller, King was of medium height, stocky, and had nondescript features. Both men idealized their mothers and were devoted to reading the Bible.[36] "The truth is," King wrote in his diary, "I see in Mr. R. precisely the same mistakes which I have heard others complain of in myself, a too great seriousness about the work in hand and too slavish adherence to a multitude of details, losing often the larger outlook."[37] It is therefore not surprising that multiple sources describe the two men as best friends.

The most important similarity between the two men for purposes of understanding the Rockefeller Plan was a shared sense of their own moral rectitude. Although he did not come from a wealthy family, King nonetheless depended upon employers' noblesse oblige to make his philosophy of industrial relations work. Businesspeople of honor supported labor, and he had the perfect example in his friend Rockefeller. "In Mackenzie King's life the Rock of Ages was Rockefeller," wrote King's early biographers, H. S. Ferns and B. Ostry. "He represented the perfect conscience both pure and omnipotent. Nothing ever shook Mackenzie King's confidence in Rockefeller."[38] What was more surprising, the two men expected CF&I workers to be forward-looking enough to join management in an effort to improve the business in which both shared an interest without recognizing that workers stood to gain little from the joint enterprise compared with stockholders such as John D. Rockefeller Jr.

Mackenzie King had impeccable liberal credentials. He had been a resident fellow at Jane Addams's Hull House and a student of liberal economist Thorstein Veblen at the University of Chicago. He organized the first Labour Department within the Canadian government and was elected a member of Parliament in 1909.[39] In 1911, King was voted out of office with the collapse of the Liberal Party coalition in national elections that year. King wanted to return to Canadian politics eventually, but he needed to earn a living and wanted to maintain his reputation as a staunch supporter of labor. When Jerome D. Greene of the Rockefeller Foundation approached King about studying the "labor problem" in general, using the situation in Colorado as a place to gather evidence, King feared doing so might ruin his pro-labor reputation.[40] Instead, this became his first job advising many American companies as they created similar, although seldom identical, ERPs in the years preceding King's ascension to the prime ministry of Canada. Since King became Canada's longest-serving prime minister, his association with Rockefeller clearly did his political career no lasting harm.[41] Rockefeller benefited from King's reputation the same way King benefited from his. While he might just have papered over the differences between business and labor at CF&I, the decision to hire King suggests that Rockefeller truly cared about fixing the problems in Colorado because King's concern for labor matters was well-known.

At this point, King's reputation as a labor supporter rested on his authorship of the 1907 Canadian Industrial Disputes Investigation Act, which made it illegal to declare a strike or a lockout in key industries such as coal mining and transportation without giving thirty days' notice so an impartial board could investigate the situation. While King was later attacked in the United States for this act, which hindered the ability of unions to strike, Canadian labor leaders welcomed the law upon its implementation since for the first time in history it represented at least some government protection of their interests.[42] The act was a clear representation of King's willingness to make compromises in the name of workers. "When I take up matters of interest to labor," he told the Industrial Relations Commission, "and I find there are certain limitations, and I can only get so much for labor, I take that. When I see a chance anywhere to advance the interest of labor, I do not forgo that chance, because I can not get all I think I ought to: I take as much as possible."[43] For King, then, half a loaf for labor was better than none.

To test his belief that labor and management shared a common interest, King had to convince Rockefeller not only to accept his ERP but to embrace the philosophy of industrial relations behind the plan as well. In the months preceding the introduction of the Rockefeller Plan, King served as Rockefeller's tutor on labor relations, explaining important issues and terminology regarding industrial relations that labor and management tended to view differently.[44] He succeeded marvelously at this task. As Raymond Fosdick later explained, "King sold [Rockefeller] on [the Industrial Representation Plan] and it became a matter of religion with Jr. It was the be-all and end-all of industrial relations."[45] In doing this, King justified management's disproportionate power in industrial relations matters by cultivating Rockefeller's preexisting belief in the righteousness of employers from his own class. King simply assumed that benevolent managers like Rockefeller would always do the right thing for their employees. Rockefeller, in turn, put his faith fully in the managers on the ground in Colorado. For example, facing strikes in both halves of CF&I's business in late 1919, Rockefeller wrote company president Jesse Welborn:

> I have an abiding faith that the justice of the company's position, its fairness in all its dealings with its men and the soundness of the principles which underlie all its relations with them will in the end gain fullest recognition and vindication. If the company can go through this entire strike period, long and trying as it might possibly be, and retain the confidence and goodwill of its employees and the respect and backing of the people of the State, a new high level will have been reached in industrial relations.[46]

It seems that Rockefeller never considered the possibility that the strikes had occurred precisely because whatever goodwill the company had once enjoyed had evaporated.

Unfortunately for the prospects of the Rockefeller Plan, many CF&I workers did not share King and Rockefeller's assumption of management's benevolence. They needed some form of protection to allow them to complain about working conditions to people who had the power to fire them, yet nothing in King's employee representation scheme provided that security. Reflecting the same misjudgment Rockefeller had made with respect to the honesty of the miners he had talked to on his 1915 trip, King did not realize that CF&I employees needed security of any kind. As he recalled in a letter to Rockefeller in 1921, "So far as I had to do with the drafting of the Plan, and so far as we both had to do with enforcing its several features, we were seeking to give expression to views of the employees as given to us after a painstaking endeavor to secure their point of view."[47] Yet they had no assurances that workers would speak to them honestly. Neither Rockefeller nor King was able to look at the distribution of power under the plan from the workers' perspective and see how it affected their willingness to express themselves freely and openly to management.

While it is easy to spot this serious flaw in King's philosophy with hindsight, the rugged individualism embodied in the Rockefeller Plan was so widely accepted in business circles during this era that it did not even appear as a value to upper-class people such as Rockefeller and

King.[48] Indeed, many businesspeople thought King's tepid scheme to redistribute limited power over working conditions to employees went too far, including a few inside the Colorado Fuel and Iron Company. Lamont Montgomery Bowers had been hired to run CF&I because he was the uncle of company director Frederick Gates. He had been successful building steamers to fetch iron ore from the Rockefellers' property on the Mesabi Range in Minnesota at the turn of the twentieth century and was therefore given the job in Colorado, with the expectation of similar success. Bowers "hadn't the slightest sense of modern labor policy," remembered Rockefeller in an interview with Fosdick. He told his biographer this because of Bowers's reaction to the ERP.[49] In 1914, when Rockefeller forwarded King's initial thoughts on an employee representation plan to CF&I management, Bowers opposed the change.[50] Prior to the Ludlow Massacre, Rockefeller had never second-guessed Bowers's judgment on labor matters, and he did not do so in this instance. However, after the strike, in December 1914, Rockefeller replaced Bowers with Jesse Welborn largely because of Bowers's continued opposition to employee representation.

King had more success bringing Welborn to the cause. In March 1915, King arrived in Colorado for the first time to try to convince Welborn and other executives to support employee representation. King spent several weekends at Welborn's house, where he rode horses, played with Welborn's children, and charmed his wife. Over time, Welborn became an important convert to the cause of employee representation.[51] In 1916 Welborn wrote King in Ottawa, "Your uniform frankness and interest shown in myself and associates immediately inspired our confidence, and 'King' is our authority now to what I think you would regard as a surprisingly large extent."[52] Rockefeller later remembered, "[W]hile [Welborn] had been brought up in the Bowers school [of industrial relations], he was much more open-minded and did his best to put the Representation Plan into effect, although it never had his complete intellectual sympathy."[53]

Despite this apparent victory, Welborn's opinion mattered little in comparison to the way workers experienced the plan on the shop floor. To implement a plan employees would use, Rockefeller and King needed cooperation not just from the president of the company but from the lower ranks of management as well. By late 1915, King thought he had converted many of the superintendents and pit bosses to "new ways of thinking and acting."⁵⁴ As Welborn later recalled, "It is, perhaps, only fair to say that some of the officials of the Company, as well as some of the mine workmen, entertained the feeling that it would be impossible or impracticable to make the principles laid down in the Industrial Plan effective. However, almost without exception both wage earners and management officials accepted the new regime established with the determination to do their utmost to make it a success."⁵⁵ Yet low-level managers might tell their superiors they believed in employee representation, but that does not mean they actually did what they said. Furthermore, even if foremen and superintendents agreed with King about the plan, this did not mean employees believed they could make complaints with impunity. As a later independent report on the plan by the Federal Council of Churches explained, "While most of the representatives interviewed declared that they were not afraid to bring up grievances, there was general agreement among them that the case is different with many rank and file who are reluctant to bring up grievances for fear of incurring the ill will of foremen and superintendents."⁵⁶ Lower-level management had no reason to coax grievances out of the rank and file because as long as workers failed to voice such complaints, it would seem as if the plan was working even though it was not. After all, they were held accountable for the company making money, not for harvesting grievances.

ROCKEFELLER VS. KING ON INDEPENDENT UNIONS

Mackenzie King's teachings did much to shape John D. Rockefeller Jr.'s philosophy of industrial relations, but the two men did not agree on all labor relations issues. Throughout their association, King showed a much greater tolerance for independent trade unions than Rockefeller did. King explained his position on the relationship between the Rockefeller Plan and independent unions to the Canadian National Industrial Conference in 1919: "I would have had no part in the [Rockefeller Plan] one way or the other if it had been even remotely intended for the purpose of fighting unions."⁵⁷ Indeed, King was the impetus behind CF&I's willingness to treat independent trade unions differently from firms such as United States Steel did.⁵⁸ However, the one condition King placed on his support for independent trade unions was that they behave in a "responsible" manner. "If unions are to have any claim," he wrote to Rockefeller, "they must prove themselves as good, not evil. Because a man or a set of men may decide to open a bank, that does not carry with it the necessity that we shall do business with it. It must prove itself and make itself worthy of our confidence. So [it is] with a union."[59] During the early days of the plan, both King and Rockefeller agreed that the United Mine Workers of America (UMWA) was not a responsible union. Therefore, the issue never arose.

If it had, there would have been tension, as King's friend had a much more complicated attitude toward independent trade unions than he did. In his January 25, 1915, statement before the U.S. Industrial Relations Commission, Rockefeller declared:

> I believe it just as proper and advantageous for labor to associate itself into organized groups for the advancement of its legitimate interests, as for capital to combine for the same object. Such associations of labor manifest themselves in promoting collective bargaining, in an effort to secure better working and living conditions, in providing machinery whereby grievances may easily and without prejudice to the individual be taken up with the management. Sometimes they provide benefit features, sometimes they seek to increase wages, but whatever their specific purpose, so long as it is to promote the well-being of the employees, having always due regard for the just interests of the employer and the public, leaving every worker free to associate himself with such groups or to work independently, as he may choose,—I favor them most heartily.[60]

To a public that had just witnessed the Great Coalfield War of 1913–1914, this statement seemed like a retreat from the Colorado industry's implacable anti-unionism inherent in that conflict, and indeed it was. However, the conditions Rockefeller placed upon the unions he approved of caused them to resemble the sham organizations so prevalent in the 1930s more than independent unions such as the UMWA. A miners' union that "had due regard for the just interests of the employer" would not have struck for recognition in 1913–1914 in the first place. More important, a union that allowed members to "work independently" faced the real danger of not being able to collect enough money in dues from cash-strapped miners to fund its operating expenses.

What is interesting about Rockefeller's attitude is that despite his support for such weak unions, he still believed employees would pick his ERP over independent organizations like the UMWA. "It seems to me that every man should have the right to decide for himself whether the Company is his best friend and champion or whether some outside organization is," he explained in a 1916 letter to CF&I executive Elmer Weitzel after the latter had proposed removing union organizers from company-owned camps. "If the company cannot convince the men that it is their best friend, that it will cooperate more zealously with them than any outside organization in safeguarding their interests and well-being and securing them the fullest protection and justice, then the men must and should ally themselves with any organization which they believe gives better assurances of such results."[61] Rockefeller had faith that workers would pick his form of employee representation over an independent trade union because the independent union would presumably only represent the interests of its own members, while the company offered representation for everyone without requiring dues. As he explained the day before he introduced the plan in 1915, "Unionism does for a group of working men—not for the great mass of working men. This plan is to protect the interests of all."[62] Rockefeller hoped to offer less-skilled, underserved employees a better union than the UMWA to represent them.

The United Mine Workers of America, particularly during the 1920s, had many problems. Under the leadership of John L. Lewis, it was hardly a democracy at this juncture. According to his most prominent biographers, "Lewis used every instrument at his command to accumulate power. Brutality, bullying, deceit, and bluff were all means Lewis used to achieve his ends."[63] For this reason, Rockefeller could have won the competition. Yet CF&I would never have offered miners the Rockefeller Plan in the first place had the UMWA not continually tried to organize its mines, and the UMWA had to work harder and be more accommodating to win miners over to its side. In the end, the vast majority of miners generally preferred the potential for improvement an independent union offered over the limitations the Rockefeller Plan imposed upon them every day. There was something about an independent union workers liked that the Rockefeller Plan could never duplicate. Most likely, it was the independence.

His recognition of the structural limitations within any company-dominated union may explain why Rockefeller's attitude toward independent unions eventually hardened into a philosophy fully distinct from that of Mackenzie King. By 1919, the younger Rockefeller no longer believed in direct competition between employee representation plans and independent unions. Rockefeller's rhetoric at President Wilson's Industrial Conference clearly illustrates the two men's differences. Wilson called the Industrial Conference to solve the enormous labor problems associated with the postwar reconstruction of the U.S. economy. Despite being an industrialist, Rockefeller served as a "public representative" to the conference. Other businesspeople served both as public members and as members who represented business. Labor leaders represented organized labor. During the conference, Rockefeller introduced a resolution that read, in part:

> RESOLVED, That this Conference recognizes and approves the principle of representation in industry under which the employees shall have an effective voice in determining their terms of employment and their working and living conditions; and be it further
>
> RESOLVED, That just what form representation shall take in each individual plant or corporation, so long as it be a method which is effective and just, is a question to be determined by the parties concerned in the light of the facts in each particular instance.[64]

Although a letter Rockefeller wrote a few months earlier clearly indicates that he accepted the possibility of trade unionism or some combination of employee representation and independent trade unionism as a solution to the labor problem, now he wanted to tilt the system to his side. Rockefeller never acknowledged the reality that when management has a hand in determining what constitutes "effective and just" representation, independent trade unions are a very unlikely outcome.[65] Yet American employers of the era would not accept even this stilted outcome. They simply were not used to being challenged in their own factories by any outside organizations. Therefore, they rejected anything that threatened their absolute control and rejected Rockefeller's suggestion out of hand. In contrast, the labor delegates to the conference found Rockefeller's proposal too weak.

Mackenzie King was more sensitive to the concerns of organized labor than Rockefeller was. For King, employee representation was a first stage that would eventually lead to recognition of responsible independent unions. "[T]his is a period of transition in which Organized Labor is bound to come in for an ever-increasing measure of recognition," he wrote Rockefeller in 1919. "The path of wisdom on the part of managements seems to be that of bringing about the necessary adjustments in the most natural way and the one least liable to lead to friction."[66] With respect to die-hard opponents of unions, King wrote, "The mind that sees only the evil, and fails to see the good, in any institution commanding the support of hundreds of thousands of men and women, is not one by which the world is likely to be greatly helped."[67]

Unfortunately for King, one of the minds that saw only evil in trade unions was another Rockefeller adviser: lawyer and CF&I director Starr Murphy. In his diary, King described a September 29, 1919, meeting with Rockefeller at which he helped prepare Rockefeller for Wilson's Industrial Conference: "I tried to show the justice of unionism particularly to Murphy, who is a poor advisor to Mr. R. from that standpoint. Under guise of fear of 'closed shop' w[oul]d not admit justice of much in Union contentions."[68] Indeed, in a letter to Rockefeller the previous April, Murphy wrote:

> I have never been able to agree with the proposition that there is no inconsistency between the Plan of Representation which we have in Colorado and the labor union idea. It seems to me that they are fundamentally opposed to one another. The Colorado Plan is based upon democracy and freedom. The labor union is based upon the idea of monopoly and coercion. The whole labor union movement is predicated upon the idea of monopoly of labor, and wherever they are able to do so they insist upon the closed shop.[69]

On this issue, Rockefeller ultimately sided with Murphy over King.

SELLING EMPLOYEE REPRESENTATION

The labor movement had originated the term "industrial democracy" in the late nineteenth century but was never exactly clear on its definition. By the end of World War I, the exact meaning of the term was up for grabs.[70] Did "industrial democracy" only mean trade union representation, or would employers play a role in determining the nature of workplace governance in a United States that was increasingly inserting democracy into areas where it had seldom been present? Would the democratic labor reforms of the World War I era continue into the postwar period even though the United States was no longer, to use President Wilson's phrase, fighting "to make the world safe for democracy"? This was the environment in which, at the urging of Mackenzie King, John D. Rockefeller Jr. began to write and speak out in favor of employee representation along the lines of the Rockefeller Plan.[71]

As the guiding force behind the most visible employee representation plan in the world, Rockefeller played a major role in the debate over what the term "industrial democracy" should mean. He became a subject of that debate before the 1913–1914 strike ended, but Rockefeller entered the debate on his own terms in early 1916 when he sent 500,000 copies of a self-published booklet entitled "The Colorado Industrial Plan" to union leaders, other businesspeople, and the press.[72] The booklet contained a copy of the plan and the text of the speech he had given in Pueblo that introduced the plan to the world. Around the same time, the *Atlantic Monthly* published an article Rockefeller wrote entitled "Labor and Capital—Partners," which clearly reflected King's influence. With respect to CF&I he explained:

> It has always been the desire and purpose of the management of the Colorado Fuel and Iron Company that its employees be treated liberally and fairly. However, it became clear that there was need of some more efficient method whereby the petty frictions of daily work might be dealt with promptly and justly, and of some machinery which, without imposing financial burdens upon the workers, would protect the rights, and encourage the expression of the wants and aspirations of the men.[73]

Most of the article was devoted to explaining the importance of interclass partnership rather than describing the Rockefeller Plan itself. The reason labor and capital should be partners, he wrote, was that "the riches available to man are practically without limit. . . . [T]he world's wealth is constantly being developed and undergoing mutation, and . . . to promote this process both Labor and Capital are indispensable. . . . [I]f they fight, the production of wealth is certain to be retarded or stopped altogether." The manner in which that wealth might be distributed held much less interest, as he argued that there will "always be injustice," but—reflecting King's emphasis on attitude—Rockefeller wrote that the "injustice of that division will always be greater if it is made in a spirit of selfishness and short-sightedness." Rockefeller went on to note, "The Colorado plan is of possible value in the State, and may prove useful elsewhere because it seeks to serve continually as a means of adjusting the daily difficulties incident to the industrial relationship."[74]

Because of this direct appeal by Rockefeller to a broader audience, speaking invitations from across the country began to pour into his New York City office. This was both a cause and an effect of Rockefeller regularly having thousands of copies of his speeches bound, printed, and distributed to potentially interested parties nationwide. For the interested parties at CF&I, the most important of his efforts to sell the concept of employee representation occurred when he returned to Colorado in 1918 (the trip on which he tried to visit the dedication of the Ludlow Massacre memorial, discussed at the beginning of Chapter 1). As on his previous trip, he took Mackenzie King with him. This time, however, his wife, Abby, also accompanied him (spending most of her time with miners' wives), and he kept the trip a secret until he arrived. The nominal purpose of the second tour was for Rockefeller to see how the money he had donated to be spent on gardens, clubhouses, and the like had been used. But in fact he had much more to accomplish than surveying the results of his gifts.

Rockefeller wanted to collect proof that the ERP had succeeded in its goal to stop labor strife. As *The Pueblo Chieftain* noted during his visit to the Minnequa Works (which had its own ERP by then), "When opportunity afforded he conversed with the workmen and seemed elated with information thus gained. He never missed a chance to ask about their welfare, and if everything about their work was satisfactory. He paid special attention to opinions regarding the employee representation plan."[75] In a private letter to his father, Rockefeller described what he heard during conversations with employee representatives. "We have spent a couple of days in Pueblo," he wrote. "[We] have met the representatives of the employees both at a luncheon where an equal number of officers were present, and also in a subsequent meeting with the representatives alone. On both of these occasions the representatives expressed their cordial endorsement of the plan of industrial representation and with full appreciation of it."[76] Having heard nothing but good things about the plan, after Rockefeller returned to New York he announced that "[i]n view . . . of the statements of the representatives, of my own observations and the results obtained during the three year period, I believe it can be said with confidence that the plan is no longer an experiment, but a proved success that based as it is on the principles of justice to all those interested in its operation, its continued success can be counted on, so long as it is carried out in the future as in the past."[77] This definitive conclusion demonstrates that, as with his earlier trip to Colorado, CF&I employees were not his only audience.

Besides learning about the success of the plan, Rockefeller used his second trip to Colorado to encourage other businesspeople to support the idea of employee representation. "What particular part shall we say that industry has in patriotism; how can industry show its patriotism?" he asked a crowd of over 1,000 people (one-third to one-half of whom were CF&I employees) in Pueblo while U.S. troops were fighting in Europe:

> It seems to me that obviously a demonstration of patriotism in industry is made through the greatest possible output; that output depends upon several things—first on industrial peace. In the past decade there has been too frequently throughout this fair land, industrial war, which labor and capital have both paid for in dollars, but now, if industrial war takes place in industry, it is no longer paid for in dollars, but in lives. . . . I think those men of the industrial classes who permit to be brought about industrial war at this time are traitors to our country, if not murderers.[78]

In other words, John D. Rockefeller Jr. directed the same kind of rhetoric at his business colleagues who opposed employee representation that most businesspeople reserved for radical unions like the Industrial Workers of the World. Speaking before the Denver Civic and Commercial Club a few days later, Rockefeller elaborated on his solution for industrial warfare using less strident language: "I am profoundly convinced that nothing will go so far toward establishing Brotherhood in industry and insuring industrial peace, both during the war and afterwards, as the general and early adoption by industry of this principle of representation, the favorable consideration of which cannot be too strongly urged upon leaders in industry."[79] For Rockefeller, the apparently friendly relations between labor and management at CF&I served as proof that his solution to the labor question had worked.

Rockefeller continued to tout employee representation after the war ended. On December 5, 1918, he spoke at the War Emergency and Reconstruction Conference of the U.S. Chamber of Commerce in Atlantic City, New Jersey. In some ways, this speech represented the same Mackenzie King–influenced ideas he had explained in Pueblo when he introduced the ERP. For example, he cited the same four parties involved in a corporation. However, this time he went much further:

> The soundest industrial policy is that which has constantly in mind the welfare of the employes as well as the making of profits, and which, when human considerations demand it, subordinates profit to welfare. . . .
>
> The day has passed when the conception of industry as chiefly a revenue-producing process can be maintained. To cling to such a conception is only to arouse antagonisms and to court trouble.[80]

Rockefeller mentioned the Colorado Fuel and Iron Company in this speech only in passing, as one example of employee representation. Nevertheless, CF&I remained important to him as the embodiment of this principle. By the time he spoke at Wilson's Industrial Conference the next year, however, Colorado did not merit a mention. By then, he had developed a complete industrial relations philosophy:

> In the early days of the development of industry, the employer and capital investor were practically one. Daily contact was had between him and his employes, who were his friends and neighbors. Any questions which arose on either side were taken up and at once readily adjusted.
>
> A feeling of genuine friendliness, mutual confidence stimulating interest in the common enterprise was the result. How different is the situation today. Because of the proportions which modern industry has attained, employers and employes are too often strangers to each other. Personal contact, so vital to the success of any enterprise, is practically unknown and naturally misunderstanding, suspicion, distrust and too often hatred have developed, bringing in their train all the industrial ills which have become far too common.[81]

The businesspeople at the conference rejected Rockefeller's suggestion that employee representation serve as a compromise to bring the competing sides in that year's industrial strife together. The labor delegates actually supported Rockefeller's resolution to that effect in its initial form.[82] His failure to win more businesspeople over to his side at the Industrial Conference probably explains why this was Rockefeller's last high-profile speech on labor matters. By the late 1920s he was too involved with major philanthropic activities such as Colonial Williamsburg and business projects like the building of Rockefeller Center to care much about what happened in Colorado.

In its early years, conservative critics attacked the Rockefeller Plan as "the altruistic hobby of a large stockholder . . . that is adhered to regardless of, and often to the detriment of, profits." At its most expensive, the plan cost management only $216,975.79 to administer, a little more than one-half of 1 percent of CF&I's revenues.[83] However, that figure ignores the effects of the plan on the firm's overall business position. Throughout the 1920s, both CF&I's coal and steel businesses suffered from fierce competition, low profits, and the cost of the $17.5 million campaign to modernize their facilities.[84] The firm needed labor peace to help it weather such difficult times. As later chapters will illustrate, rather than help achieve labor peace, the Rockefeller Plan actually became the focus of labor trouble because many CF&I employees became disenchanted with it. Had the Rockefeller Plan solved the labor question, the company might have been profitable. Instead, the plan failed to keep enough employees happy throughout its lifetime to eradicate costly strikes. Perhaps more important, the plan even failed to bring labor and management together for the good of the company when workers were on the job. As Rockefeller's consultants from the law firm Curtis, Fosdick, and Belknap explained in a private study of the plan completed in 1924, "The minutes of meetings and our discussions with employees and officials of the company indicate that while the plan has resulted in an especially high morale, there has not been a capitalization of that morale to the extent of bringing forth united efforts on the part of everyone to lower costs."[85] Rockefeller had thought that instituting an ERP would be good business, but business became too bad to save. This might explain why he eventually backed away from his early interest in labor reform. The Rockefeller Plan only aggravated an already bad situation.

Chapter Three

BETWEEN TWO EXTREMES

> Between the extreme of individual agreements on the one side, and an agreement involving recognition of unions of national and international character on the other, lies the straight acceptance of the principle of Collective Bargaining between capital and labour immediately concerned in any group of industries, and the construction of machinery which will afford opportunity of easy and constant conference between employers and employed with reference to matters of concern to both, such machinery to be constructed as a means on the one hand of preventing labour from being exploited, and on the other, of ensuring cordial cooperation which is likely to further industrial efficiency.
>
> —MACKENZIE KING TO JOHN D. ROCKEFELLER JR., AUGUST 6, 1914[1]

Since the inception of employee representation plans (ERPs) around the beginning of the twentieth century, trade unionists and others sympathetic to workers have denounced ERPs such as the one in use at Colorado Fuel and Iron (CF&I) as "company unions" because of management's tendency to control the course and result of deliberations. For example, Mother Jones wrote in her autobiography, "I told him [John D. Rockefeller Jr.] that his plan for settling industrial disputes would not work. That it was a sham and a fraud. That behind the representative of the miner was no organization so that the workers were powerless to enforce any just demand; that their demands were granted and grievances redressed still at the will of the company."[2] Arguing in favor of banning such arrangements in 1935 as part of his National Labor Relations Act (NLRA), Senator Robert Wagner claimed they made "a sham of equal bargaining power by restricting employee cooperation to a single employer unit at a time when business men [were] allowed to band together in large groups."[3] Because many firms created ERPs of questionable legitimacy during the 1930s to avoid the requirements of the NLRA, the reputation of such organizations inside and outside the labor movement had grown steadily worse over time.[4] But what if a "company union" was not a sham?

The workers who participated in the Rockefeller Plan as employee representatives had no reluctance to press for change. The meeting minutes make it immediately apparent that employee representatives in both the mines and the mill were more than willing to air their grievances. One should not automatically assume that any workers who participated in a system stacked against them to affect change were automatically compromised. This does a tremendous injustice to the CF&I employees' agency. As historian F. Darrell Munsell has explained, "In some respects, [the Rockefeller Plan] gave [miners] more than they had asked for when they had gone out on strike."[5] Over time, both miners and steelworkers extracted numerous additional concessions from management. In many instances, their unionized brethren had no comparable benefits for themselves. Yet despite this success, Rockefeller's philosophy of leading with the carrot instead of the stick did not satisfy employee demands for agency. While it took longer for employee representatives than for rank-and-file employees at both the mill and the mines to sour on the plan, a great majority of CF&I employees eventually turned against the arrangement.

Mackenzie King's professed goal of using the Rockefeller Plan to secure "minimal interference in all that pertains to conditions of employment" clearly indicates that he wanted it to replace outside trade unions.[6] As explained earlier, this was less a union-avoidance strategy than a strategy to make unions superfluous by doing what they did better than they did. The plan did not achieve that goal. However, this failure does not necessarily mean the Rockefeller Plan was no better than the sham unions created during the 1930s to avoid abiding by the NLRA. Daniel Nelson has described these organizations as effective agents of the open shop. Typically, a small group of favored employees, often directed by supervisors, would form a representation plan that met the letter of the law and that was available to thwart outside organizers. Company union officers often formed extralegal groups that disrupted union activities, attacked picket lines, assaulted organizers, and testified against pro-labor legislation.[7]

The Rockefeller Plan could not have operated in this manner because if it had, it would not have withstood the extraordinary scrutiny John D. Rockefeller Jr. invited. He wanted people to see the plan in action because he thought it offered workers better opportunities than either joining independent trade unions or having no organization at all. "Mr. Rockefeller proved his faith in his working men by giving them a weapon with which to crush him if they wished," proclaimed *Leslie's Illustrated Weekly* shortly after the plan was introduced.[8] Needless to say, this was an exaggeration. But just because CF&I workers could not crush management does not necessarily mean they would have been better off if management had never introduced the plan in the first place.

THE STRUCTURE OF THE PLAN

A resolution passed at the 1919 American Federation of Labor (AFL) convention condemned what it called "company unions" because independent unions were "the only kind of organization fitted for . . . the right to bargain collectively."[9] In fact, this resolution specifically cited the Rockefeller Plan as the model for all "company unions" at the time. The resolution, introduced by the committee that organized a strike that shut down CF&I and much of the rest of the steel industry for almost three months that same year, suggested that workers active in a "company union" could never free themselves from management's grasp. Not coincidentally, it also protected the primary constituents of the AFL's member unions—skilled craftspeople—from other workers who wanted to use employee representation to force greater equality among workers.[10] These constituents could afford an "all-or-nothing" position on independent unions because they had more skills to market than other workers, making it easier for them to organize independently.

Opponents of employee representation generally based their opposition to ERPs on their inferiority to independent trade unions. However, from the workers' perspective, an inadequate representation plan was often better than no union at all. Although most plans did not require management to relinquish control over the shop floor, even the most stifling company-dominated union was still a concession indicating that workers had some rights.[11] Employees thus had the potential to build on that concession down the line. In an industry like steelmaking, for which no viable independent trade union existed, employee representation plans were the only kind of union most workers had at the time. More important, CF&I's ERP proved extremely generous in conceding powers and offering benefits to employees, at least as far as these arrangements went. Indeed, the structure of the Rockefeller Plan ceded many rights and powers to employees that they never would have had in its absence.

One reason scholars tend to lump all "company unions" together is that the details of employee representation plans are not gripping reading. This is certainly true of the Rockefeller Plan. The first version of the plan applied only to the coal mines. A new version was drawn up when management extended the plan to the steel mill in 1916. A consolidated version of the plan (with only a few significant changes) took effect in 1921.[12] Ben M. Selekman and Mary Van Kleeck fault the language of the plan in all its versions as "not inspiring or, even, quite intelligible."[13] However, this chapter is concerned with the plan's content rather than the style in which it was written.

The Rockefeller Plan separated the company into divisions: nine in the Minnequa Works and one at each mine. Each division served as an election district for employee representatives. The plan mandated that workers elect 1 representative for every 150 wage earners, but each coal camp, no matter what its size, had at least 2 representatives. The first version of the plan divided the mines into five districts: the Trinidad District (covering properties in Las Animas County in the far southern region of Colorado), the Walsenburg District (covering properties to the north), the Canon District (covering properties around Cañon City, Colorado), the Western District (covering properties on the Western Slope of Colorado), and the Sunrise District (covering CF&I's iron mine in Wyoming). When steelworkers came under the Rockefeller Plan they became their own district, the Minnequa District, after the name of the Pueblo plant; the same rules applied for those divisions as applied to the coal camps.[14] The plan was supposed to extend for three years, but it was renewed continually until the United Mine Workers of America (UMWA) organized the mines in 1933.[15]

Under the terms of the plan, joint conferences between all employee representatives and management could be called at any time by order of the company president, but they had to be held at least once every four months. Usually, management held the joint meetings quarterly. The U.S. Department of Labor wrote about 592 ERPs in the mid-1930s and found that 40 percent of the companies had never called a meeting of their representatives or their constituents.[16] This is a clear sign that those organizations were employer-dominated shams. However, this was not the case with the Rockefeller Plan. The minutes available in the CF&I Archives make it clear that management dutifully adhered to its responsibility to keep the ERP active. This desire extended from President Jesse Welborn's willingness to hear appeals all the way down to the four joint committees active in each mine and at the mill. Employee representatives and management representatives served on these smaller bodies to discuss particular issues in greater depth. Their names indicate the subjects they addressed: Industrial Cooperation and Conciliation (later Industrial Cooperation, Conciliation and Wages), Safety and Accidents, Education and Recreation, and Sanitation, Health and Housing. The AFL resolution against "company unions" complained that companies "load their company union committees with bosses, usually to the point of a majority."[17] The Rockefeller Plan capped the representation on each joint committee at three representatives each from labor and management; furthermore, the plan stated that at the district conferences "[t]he Company's representatives shall not exceed in number the representatives of the employes." If the joint committees could not agree on a particular issue, the plan included a provision under which such questions were decided by an outside arbitrator selected by a majority of that body.

Under the Rockefeller Plan, a chairman and a secretary selected by the employee representatives ran the yearly election, which occurred every January.[18] Nominations and voting took place by secret ballot. The company provided blank ballots and boxes for the votes. The slate of candidates consisted of twice the number of spots to which a division was entitled. In other words, if a coal camp had enough employees to elect three employee representatives, the six who received the most nominations would stand for election. The plan included detailed provisions for

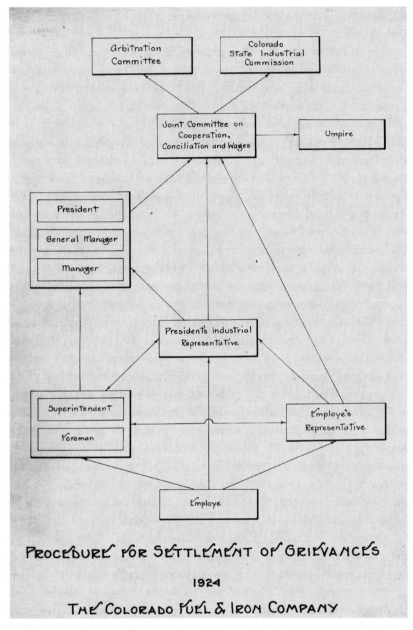

3.1. Diagram of the grievance procedure laid out under the Rockefeller Plan. The top level is the most noteworthy, since employees had the right to appeal to an outside authority, the Colorado Industrial Commission. Courtesy, Bessemer Historical Society.

recounts and election challenges, but no records indicate that employees ever used those provisions. At first, elections occurred on the job site, during working hours. Management designed the election process to demonstrate that workers were not pressured to vote one way or another. Not only did this make voting easy, but management actually encouraged it. As was the case when the miners approved the plan in 1915, however, the response was generally anemic. Perhaps because of this problem, within a few years of the plan's implementation CF&I let local units decide how best to get employees to vote. Some mines brought the ballot box to the mine entrance; others held mass meetings; still others let employees take their ballots home for the evening and drop them in the ballot box the next morning when they arrived at work.[19]

Although the plan did not mandate it at first, the company convened joint meetings once every year. They occurred no later than three weeks after the annual elections—one for the coal mines and one for the steelworks and steel-related businesses like the iron mines. In addition to the regular quarterly and annual meetings, employee representatives gained the right to call meetings on their own initiative in 1921. These were usually small parleys initiated by workers to discuss grievances. During the early years of the plan, the head of management's team for discussing grievances was called the president's industrial representative. According to President Jesse Welborn, David Griffiths, the former state mine inspector who first served in this position, was "probably better known to coal miners in the state than any other man and enjoys their confidence to a degree not equalled [sic] by any other man in the state. He has always been the friend of the mine workers, and will stand out for them and their interests."[20] In other words, Welborn wanted to leave the impression that Griffiths was a neutral mediator despite the fact that management had appointed him. Yet when Griffiths met with Mackenzie King in 1915, Griffiths told King that "[h]e had no money as an appropriation to carry on the work, and apparently no authority excepting as permitted by Mr. [Elmer] Weitzel," the superintendent of mines.[21]

Over time, management would drop even the pretense of putting an impartial man in this position. Griffiths's successors came directly from the ranks of management, their loyalty to management undoubtedly recognized by all. This development demonstrates that management's interest in fairness waned as the furor over the Ludlow Massacre gradually faded. It also reflects an inability on the part of CF&I executives to put themselves in their workers' place. A more independent industrial intermediary might have made miners and steelworkers more willing to put their faith in the system King and Rockefeller created, but it appears that no one in management ever considered anyone besides other managers for this position after Griffiths left the post.

Although it was supposed to be an equal partnership, management bore the entire cost of the plan's operation. Faced with criticism on this front, management responded, "All expenses are borne by the company since the harmony and good will between the management and employees and the furtherance of the well-being of employees and their families in the communities in which they reside are essential to the successful operation of the company's industries in an enlightened and profitable manner."[22] This is where the Rockefeller Plan most closely fit the common "company union" stereotype. As the AFL's resolution noted, "The first essential for the proper working of a genuine collective bargaining committee is that it be composed entirely as the organized workers may elect and altogether free from the company's influence."[23] This is the main issue that led to the Rockefeller Plan's difficulties with the National Labor Relations Board during the late 1930s.

Other than attending meetings, the text of the plan did not specify the exact duties of employee representatives. Therefore, the way employee representatives handled their jobs, especially grievances, evolved over time. By 1920, management was instructing new representatives as follows:

> Whenever a case is presented to him, [the representative should] satisfy his own mind as to the facts in each instance and not assume, without full investigation, that the case has as much merit as the complaintant [sic] thinks it has. . . .
> When one is dealing with a case which has merit, the representative should follow the case through until a fair decision has been reached. . . .
> He should report back to the men whom he represents as to the action taken upon individual cases and also as to the work of joint conferences and joint committees.[24]

President's Industrial Representative Warren Densmore later described his activities while serving as an employee representative in the steel mill in a similar manner:

As a representative I first investigated the merit of the case. If I felt that the case did not have merit, I endeavored to inform the complaintant [sic] of my belief in the lack of merit in the case. If he still felt that he wished to be represented in that matter, I represented his complaint as he felt. If I felt that his case did have merit, I proceeded under the procedure to remedy the condition complained about, and continued to proceed until I finally succeeded in adjusting the difference.[25]

Not all employee representatives were like Densmore. Ben Selekman described a representative in Walsenburg, John Merritt, as "a timid, shrinking fellow; he never takes the initiative, waits until the workers come to him with grievances. He has only had one grievance."[26] Perhaps more important than a representative's personality, no procedures ever developed for how rank-and-file employees should use the plan to their advantage. Other than allowing them to vote in annual elections for employee representatives, the Rockefeller Plan made no provision for employees who were not elected representatives to participate in the plan's regular operation or express any opinion about results. No matter how often management told employees to bring their complaints forward or told employee representatives to present those complaints to management, the company still held power over their livelihoods. If employees were to take the Rockefeller Plan seriously, they would have to trust management. Some employees and employee representatives proved more trusting than others.

THE RIGHTS OF THE RULED

More important than the manner of representation, the text of the ERP also spelled out employees' rights under the plan. Some of these rights were new to employees. Employee representatives gained other rights by pressing for them during the plan's operation. Examples included the right to recall representatives and the right of employees to hold meetings at any time that did not interfere with operations, the latter of which did not exist in the first version of the plan. What made all these rights novel was the fact that they were written down for the first time. Workers could now consider their situation in relation to the plan's text and explain the differences between the two in the course of their grievances. Indeed, management's decision to call for ratification by employees reinforced the notion that power in this relationship ultimately came from the ruled, not the rulers.

Another aspect of the plan that strongly differentiated it from the employer-dominated ERPs of the 1930s was the possibility of appealing grievances to an outside authority. Part III, Section 17, reads:

> In the event of the Joint Committee on Industrial Co-operation and Conciliation failing satisfactorily to adjust a difference by majority decision or by agreement on the selection of an umpire, as aforementioned, within ten days of a report to the President of the failure of the joint committee to adjust the difference, if the parties so agree, the matter shall be referred to arbitration, otherwise it shall be the subject of investigation by the State of Colorado Industrial Commission.

Colorado created the Industrial Commission in 1915, the same year CF&I implemented the Rockefeller Plan, to administer its workmen's compensation law and mediate industrial disputes. Inspired in part by Mackenzie King's Industrial Disputes Investigation Act, the commission was formed as a direct result of the Ludlow Massacre.[27] Like the government boards created during World War II, the Industrial Commission had three appointed members—including one representing industry and another representing labor. The law required both labor and management to notify the commission thirty days before any change that would affect the terms and conditions of employment at any company that had four or more employees. For example, employers had to tell the commission if they intended to reduce wages. Workers had to give notice if they intended to strike. Once given notice, the commission would send someone to the scene to investigate and encourage conversation in the hope that the parties could work out their differences before a strike occurred. Through these procedures it was hoped that disputes that culminated in violence, such as the one at Ludlow, would never again occur. At the time of its passage, the act creating the commission was the only one of its kind in the nation. Conservative critics of the commission's powers called them "novel and revolutionary in their scope and purpose."[28] Unlike other companies in Colorado, by writing the Industrial Commission into its ERP, CF&I effectively agreed in advance to obey the commission's decisions by granting employees the right to appeal to that body in all matters.[29]

In assessing ERPs as a whole from the late 1910s through the 1930s, David Brody has written, "The details varied, but the small print invariably left the final word to management."[30] As an executive at Youngstown Sheet and Tube explained, under that company's plan (and many others first established during World War I), "powers necessary to an enlightened management are not abridged."[31] The Rockefeller Plan was an exception to this general tendency. It stated, "The decision of the sole arbitrator or of the majority of the Board of Arbitration [both of which were appointed by the commission], or of the members of the State of Colorado Industrial Commission when acting as arbitrators, as the case may be, shall be final and binding upon both parties." When management included this provision in the ERP, the Industrial Commission was new and untested. It is easy to be skeptical of the magnitude of this concession, but the Colorado Industrial Commission proved to be no rubber stamp for management. The company won at least two of many disputes involving wages in the early 1920s, but the commission ruled against CF&I many times when employees appealed workmen's compensation claims to the body and again when the company repeatedly tried to lower wages during the early 1930s.[32]

Of all the rights bestowed upon CF&I workers under the Rockefeller Plan, the right to join an outside union caused the most concern in management circles. "There shall be no discrimination by the management or by any of the employees on account of membership or nonmembership in any society, fraternity or union," read Part III, Section 3 of the plan. This clause was simultaneously mundane and extraordinary. The Colorado Revised Statutes of 1908 already read, "It shall be unlawful for any employer to prevent any employee from joining any lawful union or other organization or to coerce any employee by discharging or threatening to discharge him on account of his membership in any such organization."[33] The 1913–1914 strike was, for the most part, an attempt to get coal companies to follow the law. By putting this mandate in writing, management, in effect, agreed to abide by that law. Few coal companies in Colorado made this concession before the 1930s. Some eastern companies had made similar promises but did not follow through on them.[34] Since they had this promise in writing, though, CF&I miners could use the plan as a standard by which to hold management accountable.

It might surprise modern critics who see the Rockefeller Plan as simply a union-avoidance strategy that ample evidence indicates that the company took this union nondiscrimination clause seriously. The moving force behind this acceptance was Mackenzie King. As he explained in his diary on the last day of 1914, "I told [President's Executive Assistant for Employee Representation Clarence] Hicks that if Mr. R took a position against unionism and the right of men to join the union, I myself would get out and fight against him; that it was fundamentally wrong that a man of his wealth should be placed in antagonism towards the organization of labor."[35] Rockefeller, as explained in Chapter 2, opposed unions, but he still followed King's advice. Confronted with backsliding by some senior managers, Rockefeller wrote Welborn in 1916:

> It has been heralded broadly over the land that the Colorado Fuel and Iron Company's camps were open camps, and the provisions of the Plan of Representation make this perfectly clear. This policy was not arrived at hastily or thoughtlessly, but after full and mature deliberation and with a full recognition of all that it would involve, and in its adoption the management of the Company very largely, if not entirely, acquiesced. If the Company means what it says, an open camp must be absolutely and unequivocally an open camp, without reservation or strings. Any other than the most conscientious adherence to the policy announced would lay the Company open to the charge of insincerity and would be a hundred fold more harmful to the success of the Plan than the worst condition which could be imagined as possibly resulting from a rigid adherence to this policy.[36]

This is more evidence that Rockefeller cared deeply about his public image. Public pressure, therefore, served the same function government pressure would serve during World War I. That is the reason the Rockefeller Plan predated many other ERPs. CF&I did not need the war as an incitement to act; it had the public relations legacy of Ludlow as motivation.

On an individual basis, the protection the Rockefeller Plan offered employees against being fired was even more important to employees than the right to join a union. What makes this right particularly interesting is the way workers wheedled changes to it from management over time. The original plan read, "The right to hire and discharge . . . shall be vested exclusively in the Company." The first version of the ERP included a few offenses for which employees could be discharged immediately without recourse. Management posted a list of these and other reasons for discharging someone near the entrance to each of its mines. Workers, therefore, could compare their situation to the posted rules and appeal their firing through the ERP if the two did not match. However, under the consolidated plan passed in 1921, "[T]he fairness of any action" by management to discharge an employee "shall be a proper subject for review" under the plan.[37] In other words, management allowed that its prerogative to fire employees was limited by the ERP's grievance process. As might be expected, employees frequently took advantage of this prerogative. Employee representatives also had the explicit right to appeal whenever they felt they had been discriminated against for something they had done in the course of their plan-related activities.

Over time, as Ben Selekman explained, "[T]he right of employes to review any suspension or discharge became clearly established; and further, by January, 1919, employes when reinstated were paid for any time which they had lost."[38] Whether a result of management's conciliatory attitude or the persistence of employee representatives, this development was important for steelworkers because it gave them more rights than they would have had if the plan had not existed. Indeed, winning the right to question discharges resembled the achievement of an independent labor organization more than it did that of an ERP like the Rockefeller Plan. However, no matter how much the employee representatives asserted their independence, the company and the plan could not function as well as an independent union because in the end, workers had no more power than management willingly ceded to them.

At the end of the first version of the employee representation plan was a document entitled "Memorandum of Agreement Respecting Employment, Living and Working Conditions."[39] Unlike the plan itself, which would be ratified through a vote by employees, the Memorandum of Agreement was signed by the delegates to the first company-wide conference of employee representatives on October 2, 1915. The similarity between this document and a contract with an independent union was no doubt intentional. "It is mutually understood and agreed," the memorandum read, "that in addition to the rights and privileges guaranteed the employes and the Company, in the Industrial Relations plan herewith, the following stipulations respecting employment, living and working conditions shall govern the parties." The first few sections of the memorandum outlined the circumstances under which management would run its coal towns. Section I fixed the charges for rent, lighting, and water. Section II fixed the rates employees had to pay for coal to heat their homes at "substantially [the same] cost [as that to] the Company." In Section III, CF&I agreed to continue to erect fences around the miners' houses and to haul away their garbage. In Section IV, the company committed to build clubhouses (the administration of which is discussed later). The next two provisions fixed the miners' workday at either eight or nine hours and guaranteed that they would receive their pay twice each month.

Most important, the parties agreed that wages would not be decreased but that increases would be tied to those in "competitive districts in which the Company does not conduct operations." Those districts were the eastern fields, which were largely unionized. High-wage union competitors had trouble underselling non-union ones. Yet because of this provision, the wages of CF&I miners depended upon the success of the union even though the union was not recognized in Colorado. In fact, the ERP had no jurisdiction over the wage issue under the terms of this agreement. It is therefore no surprise that Selekman and Van Kleeck found that wage changes for CF&I's miners closely matched those at competing unionized companies.[40]

WELFARE CAPITALISM, THE YMCA, AND THE "COMPANY PERIODICAL"

In addition to their other duties, the Rockefeller Plan called for representatives to serve on an "advisory board on social and industrial betterment." (This board was different from the Committee on Education and Recreation, which only had jurisdiction over the aspects of welfare capitalism covered in its name.) Another provision in the Rockefeller Plan called for the publication of a "company periodical" to serve as "a means of communication between the management, the employees and the public, concerning the policies and activities of the Company." Management called this publication the *Colorado Fuel and Iron Company Industrial Bulletin*. Most of the space touted the company's numerous and varied welfare capitalist efforts, which were tied to the employee representation plan. Welfare capitalism (called "welfare work" at the turn of the twentieth century) is the name historians have given to the programs companies instituted for their employees that were not necessary to attract and keep employees based on the demands of the labor market. CF&I was a welfare capitalism pioneer that offered a diverse array of programs as early as 1901.

Historians tend to categorize ERPs as part of a vast variety of possible paternalistic welfare capitalism arrangements, such as baseball teams, employee gardens, and stock purchase plans.[41] Upon closer examination ERPs stand out from these other efforts because only under such plans do workers presumably have some voice in the way a particular welfare capitalist effort is carried out. Even though limits exist on workers' ability to exercise their power, giving employees a chance to vote on their terms and conditions of employment is substantially different from other forms of paternalism. While many ERPs, particularly those created in response to Section 7a of the National Industrial Recovery Act of 1933 and the National Labor Relations Act of 1935, did not allow workers to exercise any independence, CF&I management actively solicited and acted upon employee input through the Rockefeller Plan. Because of these efforts, employee representatives played a major part in determining the nature of management's largesse. Management used the ERP to facilitate workers' acceptance of efforts not explicitly spelled out in Mackenzie King's founding document. The *Industrial Bulletin* was its vehicle for this effort.

Welfare capitalism at Colorado Fuel and Iron predated the introduction of the Rockefeller Plan. The company created its Sociological Department in 1901 in response to a strike that had occurred that year. Its leader was the firm's chief surgeon, Dr. Richard W. Corwin. The department made a number of positive contributions to CF&I employees' lives, especially in terms of sanitation and improved housing stock in the coal towns. Management also built water systems in some communities. With respect to health care, it built a large hospital in Pueblo (which still operates under different ownership) and sent doctors into the coal camps to give lectures with titles such as "Hygiene in Warm Weather" and "Alcoholism."[42] Perhaps the most obvious sign that these efforts were not entirely altruistic was conveyed by the magazine CF&I created to tout these activities, *Camp & Plant*. The journal's first issue announced that part of its mission was to "promote the work of the Sociological Department."[43] The publication lasted five years. Management cut back on these activities during the economic downturn in 1908 but revived them under the auspices of the ERP in the aftermath of the Ludlow Massacre.[44] In 1915, CF&I folded the Sociological Department into the plan.[45]

It is no coincidence that welfare capitalism made a comeback at CF&I in the years immediately following the 1913–1914 strike. If these programs made miners (and, by extension, steelworkers) happy, they presumably would be less likely to strike in the future. John D. Rockefeller Jr. handpicked Clarence Hicks to serve as the president's executive assistant for employee representation, basically the company executive charged with overseeing the plan. Hicks's appointment explains in large part why CF&I management farmed out many of its welfare activities to the Young Men's Christian Association (YMCA) starting that same year, as he and Rockefeller had met previously when Hicks worked at the YMCA. Having the YMCA oversee corporations' welfare activities was not uncommon, as the organization had an entire department devoted to improving industrial relations at companies in many industries. However, Rockefeller's support of the YMCA (he donated millions of dollars and served on its board of trustees) reflected his religious beliefs as well as his desire to improve the lives of his workers. The YMCA Industrial Department began investigating the role it might play at CF&I in early 1915, before the Rockefeller Plan was even in place. It extended this effort throughout the mines by the end of that year by running new clubhouses paid for by the Rockefeller family.[46] When Rockefeller chose to turn over management of CF&I welfare activities to the YMCA, he did so against the wishes of the adviser of those activities, Mackenzie King, who argued against "going in for such a building [a YMCA clubhouse] as not being part of a Co[mpany]'s business."[47] Time proved King correct. YMCAs at the mining camps cost CF&I $80,000 to build and equip and $20,000 each year to maintain. The company built the lavish Steel Works "Y" with a $500,000 gift from the Rockefeller family, completing it in 1920. The steelworks newspaper described the contents of this four-story brick building in 1927:

> Inside these four walls will be found a reading room, supplied with all current magazines; a branch of the public library, with shelving accommodation for more than 1000 books; the CF&I Technical Library; a large social room on the first floor and another on the second floor; a

lobby equal to that of any exclusive club or hotel in the country; an auditorium with a pipe organ and seating capacity for 1240 people, ventilated by forced air from fans driving 30,000 cubic feet per minute; a soda fountain; eleven billiard tables; a gymnasium, thoroughly equipped and circled by a padded running track; a tile swimming pool, 20 by 60 feet in dimension; rooms for educational purposes, eight in number, including complete equipment for cooking and sewing and other home science subjects; eight regulation bowling alleys with automatic pin spotters; a cafeteria to seat 160 people, with an ice plant and refrigerating system in conjunction; private dining rooms for luncheons and dinner parties; barber shop with three chairs; and 126 bedrooms with accommodations for 155 men.[48]

With such facilities, the YMCA had a more substantial presence at CF&I than it did in most other companies across America, including many that were much larger.[49]

To Rockefeller, the high cost of welfare capitalism seemed worth it. As he explained in a letter to his father, "I have felt that to cooperate wisely here and there with the employees of the Colorado Fuel and Iron Company, in the building of a Y.M.C.A. or a church or a bandstand, was distinctly advisable and that the good resulting in the establishing of friendly personal relations and kindly feeling was far greater than the immediate benefit brought about by the definite expenditure of a definite sum of money."[50] This shows that Rockefeller's brand of welfare capitalism closely matched the principles behind the ERP. He came to believe that continued communication through the auspices of the plan improved both welfare capitalism at CF&I and the relationship between labor and management in general. Nevertheless, the YMCA went to great pains to distance itself from management when explaining its role in the coal camps. "The Association has taken the initiative in this matter," explained Industrial Department secretary Charles R. Towson, "the Company merely granting its permission for study and report. No effort was made by the Company to enlist the Association, and no suggestion was imposed as to policy or program."[51] Management's assertion that workers could control the shape of the YMCA's programs probably accounts for this distance. Indeed, this was enough to lead UMWA leader John Lawson to endorse the YMCA's programs at CF&I mines.[52] Nevertheless, as of 1919, only 40 to 50 percent of the men had joined the "Y," probably because by that time two-thirds of the expenses of operating the clubhouses came from the miners' dues.[53] Poor, unskilled miners could not afford the charges. This explains why the joint Recreation Committee of the ERP eventually asked CF&I to take over the clubs from the YMCA because of "general inefficiency."[54] The YMCA cost miners money to join, and the committee wanted the company-owned facility to be free.

While not part of the "Y," management used the Rockefeller Plan as a vehicle for obtaining feedback about the welfare capitalist program offered at the YMCA buildings and wanted workers to credit the plan for the existence of the activities it supported. This differentiates CF&I markedly from a firm such as U.S. Steel, which simply gave its workers welfare programs as if from on high and did not care if workers participated in them at all.[55] The administration of Minnequa Hospital is a good example of how CF&I used employee feedback to influence the shape of its welfare efforts. In the January 19, 1919, *Industrial Bulletin*, Dr. Corwin wrote that "through the appointing of representatives," management gained "a more thorough understanding of the value of the medical department to employees and their dependents."[56] Presumably, it was the employees who encouraged CF&I to make all the expensive improvements (such as purchasing costly medical equipment) it instituted during the Rockefeller Plan years. Assuming this feedback really occurred, it would illustrate one of many ways workers benefited from the voice the plan provided them.

Favorable descriptions of CF&I's welfare capitalist programs filled most of the pages of the *Industrial Bulletin*. (A later company publication, the *CF&I Blast*, actually began in 1923 as the newsletter of the Steel Works "Y.") The positive stories in the *Industrial Bulletin* were intended to improve CF&I's damaged reputation among both employees and the public at large. The company sent the publication to its interested "friends" throughout the country.[57] The surviving copies of the magazine in libraries across the United States are a testament to how far the company wanted its message spread. Yet the *Industrial Bulletin* also served CF&I workers. For a company with coal towns in many remote places in Colorado and Wyoming, the magazine offered a chance to speak to all of them at once, as well as to anyone else interested in the company's efforts to put the Ludlow Massacre behind it.

It is difficult to characterize precisely how CF&I workers responded to management's welfare capitalism. For example, skilled white workers and their families tended to use the YMCA frequently; other workers rarely did so.[58] The *Industrial Bulletin*'s stories about inter-company baseball teams would have interested all kinds of workers. These players were "virtually" salaried, recalled one resident of the Walsenburg camp, and the teams were multiethnic, as was the CF&I workforce.[59] Indeed, one might argue that many workers took baseball more seriously than they did the Rockefeller Plan. When the grandstand at which the steelworkers in Pueblo played baseball burned down in 1921, one employee representative who was also a player reported, "I don't think the company has ever done anything that has been more appreciated by the employees generally than the ball park and erecting of the grandstand."[60] Assuming he was telling the truth, that would explain why management had to sell the employees on everything else it did for them through the magazine.

Criticism of management occasionally appeared in special supplements along with the minutes of ERP meetings, but it is difficult to find workers' voices—good or bad—in the magazine's regular articles. "If employe correspondents in part presented their own material," wrote the consulting firm Curtis, Fosdick, and Belknap, "there might be less ground for the impression that the Bulletin is completely controlled by the company."[61] Yet when management started a special section of the *Bulletin* for items furnished by steelworkers, employees exhibited little interest and contributed only a few articles.[62] In fact, Selekman and Van Kleeck were able to judge how little the employees participated in the creation of the *Bulletin* simply from the language used therein, writing that it bore "little resemblance to the speech of men in the mines and the steel works."[63] That was no surprise because the company wrote and published the magazine at its headquarters in Denver. When Employee Representative Andrew J. Diamond praised the ERP in January 1919, he (or perhaps the editors) felt compelled to add the phrase "I express myself voluntarily and without solicitation" to his remarks.[64] Diamond's need to make this stipulation strongly suggests why the company's message in the magazine might not have gotten through to employees.

Complaints about the *Bulletin* abound in the surviving records of Rockefeller Plan–related meetings. Employee representatives complained about the meeting minutes management distributed as special issues of the *Bulletin* as early as 1918 when Representative Thomas McGee charged:

> The Bulletin was getting absolutely rotten and . . . he didn't care to be quoted any longer; the Bulletin smoothed things over; all statements should be handed back to the men as their representatives made them. That he, Messrs. Lee, Sands and Buerman had spoken for 45 minutes in the conciliation meeting criticizing the plan, and not a word of it had been put in the Bulletin, instead the Bulletin has quoted Mr. Diamond as being in favor of the plan, and that other representatives had also spoken favorably.[65]

In May 1920, Representative Warren Densmore noted that management's long-standing policy of distributing the minutes to employee representatives in advance had been violated and that this had "resulted in misstatement of facts."[66] A management representative responded that given more time, management "would follow the practice as agreed upon," but Densmore later again complained "that the matter of sending out advanced copies, or proofs, of special committee bulletins had not been corrected, although they [the representatives] were promised this would be done."[67] Such passages demonstrate that management did its censoring later in the transcription process. Otherwise, such damning complaints would not have been recorded.

The ultimate problem with the *Industrial Bulletin* was the same as that with the plan as a whole: management retained ultimate control over its operation. The best evidence for this point comes from a long memo written by Elton Mayo of the Harvard Business School in 1928. Mayo visited Colorado that year at the behest of John D. Rockefeller Jr. to report on the progress of the ERP.[68] Near the beginning of his report on the trip, he noted "[c]ertain irregularities of procedure," which included publishing "the minutes of the meetings of employee representatives of the Steel Works as if these meetings stood on an equal footing under the Plan with meetings of the Joint Committee on Conciliation and Wages." (Mayo's reference is again to the special issues of the *Bulletin* that contained minutes of the ERP meetings for the benefit of employees.) Mayo feared that broadcasting the representatives' internal deliberations would "provoke misunderstanding in the Steel Works in addition to giving rise to a false conception of the status of the employee representative meeting under the Plan."[69] The *Bulletin*, in other words, existed for management to promote its message about its benevolent welfare capitalism to its employees, not for the employee representatives' message to reach their constituents.

To keep employees from reading negative messages, "[T]he minutes are extensively 'jerrymandered' [sic]," Mayo explained. "[S]peeches deemed unwise are taken out, occasionally statements that have not been made are put in."[70] In fact, management received transcripts of the employee representatives' private meetings.[71] Therefore, representatives could never express themselves freely without fear that their opinions would displease management. If workers accepted management's assertion that CF&I's version of industrial democracy resembled government in general, they would have understood that they did not truly enjoy freedom of speech. At a time when the United States had just fought a war "to make the world safe for democracy," they would have easily seen such an obvious discrepancy.

A MODEL ERP

World War I had a dramatic effect on the number of employee representation plans operating in the United States. Under pressure from the Wilson administration's National War Labor Board (NWLB) and especially its chair, former president William Howard Taft, 120 American firms introduced shop committees (another name for ERPs) during the war to satisfy demands that they "bargain" with employees to prevent strikes from disrupting the war effort. Another 125 firms introduced such plans voluntarily. Most employers strongly preferred these plans to any independent union.[72] The NWLB advocated the use of employee representation to facilitate communication between employers and employees in an effort to avoid strikes that might damage the war effort. National War Labor Board decisions were not binding, so Taft and his allies on the NWLB had to use moral suasion and the threat of bad publicity to pressure recalcitrant employers into bargaining with these ERPs or industrial councils, as some companies called them. No matter what the name, the existence of these organizations precluded bargaining with independent unions, but it did not preclude workers from organizing inside ERPs.[73]

NWLB officials wanted to support union organization of war plants but did not feel comfortable imposing unions on previously non-union firms. Employee representation was designed as an intermediate step toward that goal. While some business leaders happily accepted government-inspired ERPs precisely because they were not the outside unions labor wanted, in many factories the organizations nonetheless became vehicles for independent union activity.[74] Historian Joseph McCartin has used the archival record of the federal government's involvement in industrial relations at plants across the country to show how NWLB-imposed shop committees became vehicles for union organizing. "Every record of NWLB shop committees indicates that unionists took control of these committees," he has written. "Moreover, once they controlled shop committees, union militants frequently used them as front organizations through which to present union demands to open-shop employers."[75]

Labor had an opportunity to use shop committees as a vehicle for organization because the NWLB faced the same kinds of problems Colorado Fuel and Iron did when it tried to determine the nature of the representation plans it imposed on specific employers. In its statements and decisions, the board simultaneously ensured that employers did not have to recognize unions and that workers had the right to bargain collectively through their chosen representatives.[76] How could these two promises be reconciled? As long as the government was directly involved in a firm's industrial relations, it balanced out the inherent unequal distribution of power between labor and management so that workers could achieve important victories. After the war, when the government lost its will to enforce the edicts of the NWLB, the situation returned to normal, and many companies drove independent unions from their shops. Companies that retained their ERPs after the war ended invariably changed their plans so workers could not take them over as they had during the war. These limitations forestalled the potential for true industrial democracy.

Companies that gave employee representation a chance during the war, whether voluntarily or otherwise, experienced different results. The NWLB's general policy was that "if an employer is to deal collectively with representative groups of his employes, it is essential that the desires of the employes should be known and be given at least equal weight with those of the employers."[77] For this reason, when shop committees remained under the direct supervision of the NWLB, the committees and the Rockefeller Plan had much in common. Indeed, the NWLB and other government agencies with similar jurisdiction and the power to do so employed large sections of the Rockefeller Plan when devising employee representation arrangements for recalcitrant businesses.[78] Therefore, as Gary Dean Best has concluded, "The Rockefeller Plan accounted for all but a handful of the over 200 plans in existence in 1919."[79] Nevertheless, all Rockefeller Plan–inspired ERPs were not the same. Usually, when the NWLB imposed an ERP on a factory, it offered labor unions more potential to grow because such arrangements were administered in a more evenhanded manner than were those over which the board exerted no influence.

After World War I ended, some businesspeople foresaw employee representation plans as the future of industrial relations in America. Economist Sumner Slichter, writing in 1929, noted, "*In short, every aspect of the post-war labor situation might be expected to cause employers to abandon their newly-acquired interest in labor's goodwill and to revert to pre-war labor policies*. And yet this has not happened. On the contrary, the efforts to gain labor's goodwill have grown" (emphasis in original).[80] Slichter was right in the sense that many employers retained the ERPs, but he was more wrong than right. Enough employees did abandon their ERPs after the war to receive a sanction from the National Association of Manufacturers, which feared such action would create trouble for U.S. industry in the long run.[81] As Sherman Rogers of *The Outlook* put it, some plants "never tried out [employee representation] fairly. Those that did entered into the arrangement with all the enthusiasm of cold-storage oysters, and of course the results were very unfavorable. . . . It is quite natural that the system failed under those conditions."[82] In short, the problem was that these companies lacked the leadership of people with the same degree of enthusiasm for employee representation that John D. Rockefeller Jr. displayed.

Because of this backsliding, any idealism regarding employee representation plans that had existed earlier in the decade had mostly disappeared by the late 1920s. Even ERPs that survived the war generally atrophied over time, as companies forgot about earlier claims of industrial democracy and increasingly operated as anti-union devices. About 400 firms nationwide still had ERPs in 1928, according to Sanford Jacoby. That might sound impressive, but Jacoby has qualified the ERPs' effectiveness:

> [C]ompanies rarely allowed representatives to meet on their own and imposed restrictions on the topics that could be discussed at meetings between representatives and management. As a result, important issues were generally ignored. Instead, these "company unions"

concerned themselves with such trivial and uncontroversial matters as plant safety and sanitation, details of welfare plans, and the purchase of candy for Christmas parties.[83]

At firms that still aspired to operate in a democratic fashion, these organizations sometimes became the building blocks for truly independent unions.[84] The Rockefeller Plan at CF&I followed both these patterns—the ERP in the mine gave way to the UMWA, while the mill plan lost its relevance for the vast majority of steelworkers. Because of the availability of the minutes of ERP-related meetings, it is possible to offer great detail on exactly how this happened in each segment of CF&I's business. The minutes document the increasing frustration workers felt as their demands to management met with immovable resistance. That frustration proved to be the best possible advertisement for independent unions the workers would ever need.

Chapter Four

DIVISIONS IN THE RANKS

> Under the Industrial Representation Plan, you are like a general without an army.
> —UNNAMED CF&I EMPLOYEE REPRESENTATIVE QUOTED IN BEN M. SELEKMAN, 1924[1]

Andrew J. Diamond was born in Joliet, Illinois, on June 27, 1877. In 1896 he went to work in the rod mill at Illinois Steel Company, which became part of the giant U.S. Steel Corporation in 1901. Diamond left Illinois Steel to take a series of jobs at small independent mills throughout the Midwest and then began a three-year stint at Colorado Fuel and Iron (CF&I) in 1903. Through working these jobs, he became a member of the Amalgamated Association of Iron, Steel and Tin Plate Workers, the only independent union representing workers in the steel industry at the time. In 1913, Diamond returned to CF&I to work in the rod mill. When CF&I steelworkers began representation under the Rockefeller Plan in 1916, he signed his name to the Memorandum of Agreement since his colleagues had elected him a representative for his division of the mill. He served for four years, took a brief break, and then returned to service as a representative under the employee representative plan (ERP). In 1925, Diamond became chair of the employee representatives at the Minnequa Works, a position he held until the National Labor Relations Board ordered management to dismantle the ERP in 1942.[2]

His daughter remembers Diamond as a friendly man, a trait visible in pictures of him that survive in which he is always smiling. She also remembers him as a gentleman of the old school—popular, courteous to women, always willing to meet with his constituents at the mill or even at home if necessary.[3] Chubby-cheeked as a youth, he grew portlier with age, perhaps because as his role at the mill became more important, he spent less time actually making steel. Despite his long tenure with the company, Diamond was never promoted to foreman, even though he had supervisory functions in his regular job. Instead, he gained power and prestige at the plant through the Rockefeller Plan. While others faced irregular employment during the Depression, Diamond worked regularly because he had become the lynchpin of the employee representation system, the man who kept the ERP from breaking apart under the tensions that threatened it from its inception.

Diamond believed in the Rockefeller Plan even before it made him important at CF&I. "In the many years' previous experience I have had in the steel mills the country over," he wrote in 1919, "I saw and learned of nothing that could compare with the Plan as an arrangement sincerely and effectively to promote good working conditions for the employees and like living conditions for them and their families and welfare work of all kinds."[4] He made this statement in the *Industrial Bulletin* despite his experience as a member of an independent trade union. In response to a 1924 circular letter from management, Diamond wrote, "After eight years of its operation, the Plan speaks for itself as an unqualified success. Each year there is a stronger and better feeling between the employees and management."[5] When Elton Mayo visited Colorado in 1928, he met with Diamond in Pueblo. In a long memo to Rockefeller's office, Mayo explained that Diamond was given three or four days off each week. He used that time to help mediate grievances, even though he was not paid for it. Diamond said he did this "in the best interests of the men and the company." For this reason, Mayo concluded that Diamond's "loyalty and sincerity are unquestionable."[6] Indeed, wrote Mayo, "[O]ne of Mr. Diamond's preoccupations . . . is fear lest the Plan may be altered or abandoned was indicated by almost every comment he made."[7]

Other CF&I employees were less enamored with the Rockefeller Plan. When sociologist Eric Margolis interviewed former miners in the

4.1. *Andrew J. Diamond, undated photograph. Courtesy, Rose Ledbetter.*

late 1970s and early 1980s, he found them still divided over the ERP. Some were grateful for the improvements it brought to living conditions; others resented the fact that management controlled the way the plan operated. One former CF&I employee expressed both sentiments at different times in the same interview.[8] Other miners who worked alongside the writer Powers Hapgood at Frederick in the early 1920s did not even know the plan existed.[9] How could this be? Race played a major part in their ignorance of the plan. Researchers for the Federal Council of Churches found that at one mine at which most employees were Mexicans and Mexican Americans, "the representatives said they 'didn't know' how the workers liked the C.F.&I. plan" and "that 'nothing much' was done through it in adjusting grievances."[10] Mexican miners who had only recently arrived in the United States usually spoke no English. Very few plan-related documents were translated into Spanish, and there is no evidence that CF&I had anyone translate what was said at meetings or even the meeting minutes.

The difference in participation rates in the plan by those of different races, therefore, should not be surprising. Naturally, the plan affected workers of different skill levels and different races in varying ways, yet it made no accommodations for such differences. Andrew Diamond—Caucasian, American-born, and highly skilled—had more in common with low-level bosses than he did with many of his fellow steelworkers. Diamond was also different from about half of the CF&I workforce during the 1920s because he was a steelworker rather than a miner. In addition to the different kinds of skills needed in the two sides of the business, the miners had a union they could turn to for help during the existence of the Rockefeller Plan (even if management refused to bargain with that organization). Steelworkers, on the other hand, with the exception of about a year between 1918 and the 1919 strike, had no effective union in their industry. Yet despite these differences, the eventual response of most CF&I employees in both sides of the company was the same: apathy. That apathy started among rank-and-file employees and steelworkers, many of whom were Mexican or Mexican American, and eventually spread to the employee representatives, even if it never reached Andrew Diamond.

DIVERSITY AND ORGANIZATION

"Most westerners saw the rapid development of the West as their greatest triumph," historian Richard White suggested. "They calculated their achievement in the terms that development itself mandated: tons of ore, miles of rail, bushels of wheat, heads of cattle. They converted all of these into dollars, and there is no doubt that their achievement was great."[11] Even though westerners had mined coal as early as the 1850s, the absence of the word "coal" from White's list of regional accomplishments is telling. Coal mining in the West has never drawn the same amount of attention from historians as has coal mining in the East, perhaps because the story of the coal industry is less dramatic than the story of gold and silver mining. White's failure to identify steel as a product of western development is more understandable. For the first forty years of the company's history, Colorado Fuel and Iron operated the only steel mill west of the Mississippi. If CF&I's competitors wanted to sell steel in the West, they had to ship it in from Chicago and points farther east. What made CF&I a distinctly western company was not its products but its workforce. Starting with a nationwide spike in the number of immigrants during the 1880s, the company drew on the same cohort of mostly Southern and Eastern European immigrants tapped by its eastern mining and steelmaking competitors. But unlike eastern firms, the company also drew on the unique mix of races and ethnicities in the American West, especially the Mexicans and Mexican Americans who lived in and around southern Colorado.

As late as the 1890s, the majority of Colorado coal miners were Americans or immigrants of English, Scottish, Welsh, or Irish descent. These workers' tendency to go on strike contributed to the growth in the number of Southern and Eastern European immigrants in the state after 1880. African Americans entered Colorado mines for the first time during the early twentieth century. By 1923, CF&I employees came from fifty-four different nations and lived in twenty separate communities.[12] By that time, the largest minority group in both the mines and the mill were Mexicans and Mexican Americans.[13] Pushed out by the revolution in their home country and pulled in by the prospect of better-paying jobs north of the border, Mexican immigrants began to enter the region in large numbers before World War I. After a brief lull during the economic downturn in 1921, the number of Mexicans crossing into the United States accelerated again in 1923, quickly surpassing the wartime rate even though much of that immigration was now technically illegal.[14] Between 1920 and 1930, when the number of foreign migrants to Colorado dropped dramatically overall, the state's Mexican population increased by 400 percent.[15] Nationwide immigration restrictions on Europeans, which Congress first imposed during the 1920s, ensured that CF&I would continue to deal with these non-European immigrants for years to come.

In 1916, Mexicans and Mexican Americans constituted 607 of the 2,601 Colorado Fuel and Iron employees in the Trinidad District, the company's largest concentration of mines.[16] By 1924 they constituted 714 of 2,090 miners in the same district.[17] This increase, from 23 to 34 percent of the workforce, is indicative of changes throughout the company during this time. On August 1, 1918, approximately 20 percent of CF&I's miners were Mexican or Mexican American.[18] After the strikes in 1919 and 1921, as many as 60 percent of its new hires in both the mines and the mill were people of Spanish or Mexican ancestry.[19] While Mexican and Mexican American workers made up the largest ethnic group in CF&I's employ, they were not the only significant minority. Other groups included Italians (still the largest minority in 1916 until surpassed by Mexicans and Mexican Americans), Germans, Austrians, Greeks, Poles, Swedes, Japanese, and Chinese. Spurred by previous efforts by Colorado coal companies to inspire divisions between workers along racial and ethnic lines and CF&I management's own obsession with the race and ethnicity of its workforce, Margolis has written that miners became "hyper-conscious of ethnicity." He still detected that attitude when he interviewed former miners decades later.[20] Yet to produce profits, management had to turn this multiethnic, multiracial workforce into an efficient, well-coordinated, harmonious body. Racial tension made achieving this goal very difficult.

In the early twentieth century, Coloradoans generally described the state's Hispanic population as either "Spanish," "Spanish American," or "Mexican." Paul S. Taylor found these terms in common usage in his comprehensive 1929 study of Mexican sugar beet workers in Colorado's Platte River Valley: "The term Spanish American as used among the laboring classes is confined almost entirely to the Spanish or Indo-Spanish descendants of the inhabitants of Southern Colorado and New Mexico prior to 1848 when they became part of the United States. . . .

4.2. *Float from a CF&I "Field Day" parade, ca. 1920. Courtesy, Bessemer Historical Society.*

Most Americans refer to them all as 'Mexicans' quite indiscriminately."[21] The term actually originated with Mexican Americans themselves, who adopted it during the World War I years to simultaneously differentiate themselves from recent Mexican immigrants of suspect loyalty and to facilitate their own classification as white.[22] Miners and steelworkers, who interacted with this population at work every day, used this classification more often than other Coloradoans did, at least during the years before the 1927 strike. They generally referred to these less Americanized Mexicans as "Old Mexico Mexicans," meaning those from the country of Mexico. "Spanish Americans" were people of Mexican descent who had been American for generations. In Colorado, Mexican Americans made up 57.1 percent of the "Mexican" population in the 1930 census. Taylor documented considerable tensions between Mexicans and Mexican Americans during this era.[23] Light-skinned people of Mexican descent participated in the construction of the racial category "Spanish American" largely to separate themselves from others with the same heritage.[24] Therefore, the emergence of this category is itself a reflection of that tension. This conflict as well as the even more significant white/non-white division had to be breached for the Rockefeller Plan to function as its proponents imagined.

The unequal participation in the ERP by different racial and ethnic groups exacerbated, rather than improved, the tension between "Mexicans" and whites in the mines and at the mill. "The miners' representatives included white men, negroes, Italians, Austrians, Poles, Hungarians, Lithuanians, Scotch men—in all nearly twenty nationalities," reported *The Denver Post* when John D. Rockefeller Jr. announced the plan in 1915.[25] Considering the already significant percentage of Mexican and Mexican American employees at the time, the failure to include "Mexican" in that list is striking. Table A2.1 (see Appendix 2) charts the names, nationalities, first year with the company (or first year as a representative), and occupation of every employee representative in the coal mines for whom information is available, from the first year of the plan in 1915 through 1928 (the last year the *Industrial Bulletin* published a complete list of representatives). It is based primarily on personnel cards in the CF&I Archives. Those cards are incomplete, so not every representative's information is available. Nevertheless, trends regarding the representatives' race/ethnicity, seniority, and skill level are clearly discernible.[26] Approximately twenty-seven Mexican and/or Mexican American miners served as employee representatives during this period, a figure not even close to the large percentage of these workers employed by CF&I, especially after 1920. Furthermore, while many white representatives were reelected to office year after year, comparatively few Mexican or Mexican American representatives had any longevity in those positions. Many only served for one year.

Table A2.1 also illustrates that many of the jobs held by employee representatives in the coal mines were higher-level positions than those who actually dug coal. In general, there were two classes of workers in coal mines: laborers paid by the ton and those paid by the time they worked. Electrician's helpers, drivers, and lamp men were known as shift workers.[27] They were paid by the hour. The three classes of bosses in the mines—the mine foremen, superintendents, and fire bosses (who acted as safety inspectors)—were salaried.[28] Many of the employee representatives in the coal mines served in these top positions. As Ben Selekman wrote with respect to the Walsenburg District in 1919, "[N]ot a single representative was a miner who was a coal digger. They were either outside men, Fire Bosses, outside foremen, or inside Company men. The explanation of this was that miners are mainly union men and don't take an interest in the plan."[29] As discussed in previous chapters, because so many employee representatives served in lower-level management positions, rank-and-file miners were often reluctant to file grievances. Miner Jose Villa, referring to the situation in Walsenburg, explained to the Colorado Industrial Commission in 1928: "The representative of the men there in that mine was the driver boss. . . . [H]e is the boss of the Company and for that reason he cannot be in favor of the working man in that mine."[30] Workers who saw the system stacked against them had no reason to take management's promise of industrial democracy seriously.

Most miners were laborers who did piece work and were therefore paid for each ton of coal they dug out with their picks. In 1913, only 25 percent of the coal produced in Colorado was dug by machine, compared with a national average of 51 percent. In 1923, the Colorado figure rose to only 47 percent, compared with 67 percent nationwide.[31] Therefore, working as a digger in Colorado was a particularly difficult job. Diggers competed for the foreman's favor in an effort to get the best positions in the mine, such as those in which the coal had the least slate to pick out or those in which miners did not have to stoop over to dig. Some miners later accused miners of different ethnic groups of regularly paying off foremen to get the best assignments.[32] Upward mobility for Mexican miners and other non-Americanized immigrants was also limited by the requirements of state law. Starting in 1913, miners who wanted to be fire bosses or foremen had to pass a written examination. Besides the obvious language barrier, a lack of education limited the achievements of non–native-born Americans because the questions required a "thorough understanding of the scientific principles of coal mining."[33] It is not difficult to imagine the discriminatory effects of that requirement.

Another problem all miners faced was intermittent employment, working only a few days a week or even less. As long as they were not salaried, working fewer hours lowered miners' take-home pay. For decades, mining companies throughout the United States had shown a tendency to underemploy unskilled laborers, in part because of the ease with which miners could be replaced and in part because of a tendency to close mines entirely during periods of slack demand. As a result of these policies,

Table 4.1. Turnover at Minnequa Works, 1922–1929

Year	New Employees (Percentage)	Old Employees Hired (Percentage)	Rehired Total	Annual Turnover (Percentage)
1922	3,668 (48)	4,004 (52)	7,672	Blank
1923	1,853 (34)	3,567 (66)	5,420	120
1924	932 (24)	3,069 (76)	4,001	87
1925	991 (24)	3,056 (76)	4,047	80
1926	1,449 (41)	2,100 (59)	3,549	78
1927	584 (20)	2,188 (80)	2,772	72
1928	215 (9)	2,154 (91)	2,369	66
1929	282 (16)	1,546 (84)	1,828	54

Employment, January 24, 1930, 1930 Minnequa Steel Works Annual Joint Meeting Minutes, ERP Materials, CF&I Archives.

Colorado miners of all races and ethnicities had demonstrated extraordinary mobility since at least the turn of the twentieth century. Most had crossed an ocean to be in the United States and were willing to keep moving to find jobs that provided stability.[34] In the seventeen months between April 1923 and August 20, 1924, the turnover rate for CF&I mine workers was 80 percent.[35] The situation was no better in the steelworks. Table 4.1 shows the turnover rate at the steelworks from 1922 through 1929. While the annual percentage dropped during this period, it shot up again during the Depression when demand for CF&I steel plummeted.

CF&I recognized that its employees were unhappy with this situation as early as the 1913–1914 strike. In a letter written in October 1914, President Jesse Welborn cited "regularity of work" as the first problem the forthcoming ERP should rectify.[36] Yet intermittent employment remained an issue at CF&I throughout the life of the plan. The period from 1921 through the late days of the Depression was slack for the entire coal mining industry. CF&I miners worked an average of only 184 days each year between 1921 and 1927.[37] In 1924, the industrial relations consulting firm of Curtis, Fosdick, and Belknap advised management that "[s]ome form of assurance of regular work, or, in the absence of that, some form of unemployment insurance is a fundamental requirement to industrial peace."[38] Nevertheless, miners in northern Colorado (where CF&I had no mines) worked on average 30 days more in 1926 than those in Huerfano County (where CF&I owned virtually all the mines).[39] Any serious attempt to make all workers partners with management would have needed to address this problem, but CF&I never took intermittent employment as seriously as its workers did.

It is only logical that those with the least-skilled jobs faced the most intermittency, and those with the least-skilled jobs at CF&I were Mexicans and Mexican Americans. For example, at the Pictou Mine in September 1928, American and Italian miners worked an average of 20 days that month; the average was 15 for Mexican miners. At the Ideal Mine in March 1928, the average Italian miner worked 20.8 days that month; the average Mexican worked 15.6 days.[40] Even the most successful Mexican and Mexican American miners faced the same intermittent employment as miners of other races and ethnicities. Frequently, management closed mines for months at a time or laid off miners at open mines at various times in any given year. The company also closed some mines permanently during the plan's history, leaving miners in a lurch (especially if they owned property in the area) unless they received a transfer.[41] While mine shutdowns should have hurt miners of all races equally, management protected some white workers at the expense of Mexicans and Mexican Americans by transferring them to jobs at other mines. "We undertake to find places for the older employees and married men working at the mines that are closed, either temporarily or permanently," declared President Welborn in 1928. "[T]hose men will be placed as fast as possible at our other mines and, in turn, they will displace newer employees and single men who find it easier to secure employment at other occupations." Asked specifically about Mexicans, he explained that only the most productive Mexican Americans worked steadily throughout the year.[42] Despite occasional rhetoric to the contrary, management ultimately decided that intermittent employment was not a serious problem if it meant CF&I would have to pay higher wages to solve it. Therefore, the company never acted on the issue.

Similar divisions along racial and class lines existed in the steel mill. In a 1920 letter to John D. Rockefeller Jr., Welborn attributed a drop in production at the Minnequa Works "in large part to the poor quality of common labor, which is now composed very largely of Mexicans who are physically small and indolent, while prior to this year the large rugged Europeans predominated in that class of labor."[43] This is classic language for the justification of a racial hierarchy designed to keep these steelworkers in inferior positions in the mill.[44] Table A2.2 (see Appendix

2) charts the employee representatives in every division of the steel mill who served from the first year of the plan (1916) through 1928, as listed in the *Industrial Bulletin*. The table was assembled in the same manner as Table A2.1.

The most striking aspect of Table A2.2 is the paucity of Hispanic names, even compared with the number who served in the mines. One Mexican representative whose racial identity can be determined from his personnel card, Jose Castaneda, was elected in 1922, shortly after the number of Mexicans and Mexican Americans working for the company began to surge. His personnel card shows that he changed jobs eleven times over his career, five times as a result of being laid off. His job when elected employee representative was shoveling coal ash out of the boilers, one of the least skilled jobs in the plant. He only served one year in the position. In contrast, other employee representatives with identifiable jobs who also served that year included a blacksmith, two roll turners, a bricklayer, and a foreman—all skilled positions. The detailed survey of employee representatives conducted in 1924 revealed that the average length of service with the company was twelve and a half years. With such long tenures, employee representatives achieved the most highly skilled jobs in the plant, such as machinist or crane operator.[45] By keeping less-skilled workers who were mostly Mexicans and Mexican Americans from exercising power through the plan, white workers could protect their privileged positions.

While skilled workers no longer dominated steel mills like the Minnequa Works the way they had in the late nineteenth century, divisions in different departments still existed between those with skilled jobs and those with jobs that required virtually no skill. At open hearth mills like the Minnequa Works, skilled jobs made up 36.3 percent of positions in 1910.[46] Workers in the skilled jobs—lever operators, furnace operators, crane operators—used their positions as employee representatives to protect their jobs against the wage compression caused by steel companies' continual efforts to improve productivity.[47] Longtime CF&I steelworker James Jones later recalled the situation in the early days of the plan at the mill: "The first year and a half I thought I would let the younger men go as representatives, but when I heard their talk I said, 'Step aside and I'll go as representative. . . . I see what's wrong, you're trying to use this for your own benefit and not for the good of the men or the plan, as it was intended.'" (Jones served only one year as an employee representative.)[48] The racial division at CF&I closely overlapped the division of skill in the plant (and hence also of salary). Elton Mayo reported that the workforce at the Minnequa Works was 50 percent native-born white and 40 percent Mexican when he visited in 1928, yet no Mexicans held skilled jobs. "The great majority of them were unskilled laborers," he wrote in his report.[49] (This is approximately twice the percentage of native-born Americans who worked at steel mills in the East during this era.)[50] Most employee representatives were white and native-born because they had the power and the numbers to dominate the election process. As a result, they used that power to keep the benefits of the plan away from Mexicans and Mexican Americans.

LIFE AT THE BOTTOM

Colorado Fuel and Iron first hired Valentio Martinez on April 20, 1922. His job at the coke plant near the Minnequa Works was to clean up the washer's gallery, and he was mentioned in the minutes of the ERP only because he died in an accident at the plant on September 28, 1927. The minutes mention that he worked at CF&I only "intermittently."[51] This makes him representative of many Mexican and Mexican American workers there. While it is impossible to tell if Martinez spoke English, the description on his personnel card as "Mexican" (as opposed to Spanish or Spanish American) and his job at the steel mill strongly suggest that he did not. If true, this probably contributed to his fatal mishap. If a miner or steelworker did not fully understand the safety instructions he received, he was much more likely to suffer an accident. In 1912, for example, the state mine inspector reported that only 22.45 percent of miners in Colorado who suffered a fatal accident were American or of English descent. He made no mention as to whether the rest of the accident victims spoke English, but as with the Martinez case, considering the massive immigration occurring at the time, that figure is highly suggestive.[52]

The Martinez case is suggestive of another important factor for those at the bottom of the economic ladder at CF&I: the intermittent tenure of Mexicans and Mexican Americans effectively prevented them from advancement. A 1921 study of Mexican steelworkers in Pueblo by Roy E. Dickerson of the YMCA's Industrial Department found a startling

4.3. *Personnel card of Valentio (or Valente) Martinez, killed in an accident at CF&I's Minnequa Works in 1927. This side of the card indicates how irregular employment was for most Mexican and Mexican American workers during the 1920s because he entered and left employment with CF&I so many times over just five years. Courtesy, Bessemer Historical Society.*

variation in the number of Mexicans newly employed at the Minnequa Works on a month-by-month basis in the years preceding his visit (Table 4.2). His findings indicate both the seasonal nature of work at the mill and further evidence of the huge upsurge in the number of Mexican laborers hired by CF&I after the 1919 strike ended in early 1920. Equally important, however, is Dickerson's later finding that the numbers of whites and Mexicans who left employment at the steelworks during much of 1920 were almost identical. In other words, the number of Mexicans who left represented a much higher percentage of total Mexican workers than was the case for whites. This strongly suggests that most Mexican and Mexican American workers left CF&I's employment involuntarily.[53] Dickerson's simultaneous study of "Mexican Industrial Life" at CF&I coal camps reported managers' high praise for the skilled "Spanish Americans." He still noted, however, that "[t]he estimates of superintendents as to the reliability and desirability of Mexican labor mention one weakness, namely that the Mexican employee is apt to be more or less temporary."[54]

Table 4.2. Newly hired Mexicans at Minnequa Works, 1916–1920

	1916	1917	1918	1919	1920	Total
January	25	58	73	220	156*	532
February	35	52	163	34	469	753
March	46	121	245	35	387	834
April	97	134	274	156	450	1,111
May	111	118	255	117	405	1,006
June	88	167	168	196	289	908
July	145	253	261	235	431	1,325
August	91	227	256	205	328	1,107
September	157	212	137	157	317	980
October	55	117	191	0*	203	566
November	180	81	465	4*	375	1,105
December	106	162	250	30*	11†	559
Total	1,136	1,702	2,738	1,389	3,821	10,786

Notes:
* Strike on during these months.
† Company began to lay off men.

Source: Roy E. Dickerson, "SURVEY: Mexican Industrial 9 and Boy Life Pueblo, Colorado," May 13–18, 1921, Elton Mayo Papers, Box 3b, Folder 18.

The implication was that the workers themselves, rather than CF&I's employment policies, were to blame for the nature of their employment.

In fact, CF&I management not only understood the effect of its policies on Mexican and Mexican American workers but deliberately cultivated these groups' itinerancy. As management's Elmer Weitzel explained in a 1919 letter to the head of a congressional committee investigating coal prices, "The supply of labor at the mines adjusts itself to the demands throughout the seasons of the year and generally speaking only such men as are able to make good wages remain at the mines permanently."[55] Weitzel mentioned that the most likely supplemental work for the nonpermanent workers was picking sugar beets, a job almost exclusively filled by Mexicans and Mexican Americans. CF&I's management undoubtedly knew this, as it collaborated closely with the Great Western Sugar Company—the largest company in Colorado—to make sure both firms had enough labor when they needed it most. Many Mexican workers had first been drawn to Colorado by recruiters for Great Western Sugar, which needed labor to block, thin, and harvest its sugar beets for approximately fifty days each summer. Great Western encouraged coal companies like CF&I to hire these same workers in the winter (when demand for coal was highest) because if the workers were year-round residents of Colorado, other sugar companies outside Colorado could not compete for their labor. The coal companies agreed because Great Western used a large amount of their coal and also because the arrangement helped guarantee a steady supply of workers. Therefore, many Mexican and Mexican American workers at both CF&I's mine and mill found themselves in a revolving pattern of leaving one employer for another, seldom staying with CF&I long enough to reap whatever gains they might have bargained for through the Rockefeller Plan.[56]

Mexican and Mexican American employees in both the mines and the mill also had to deal with their coworkers' racism. As Dickerson noted, "Racial illwill [sic] is a two-sided affair and must be thought of in terms of educating the Americans as well as the Mexicans." While Dickerson believed the racial situation was positive at the coal camps he visited, he still noted that "Mexicans display considerable reserve and do not seem to feel free to meet and mingle with Americans and others excepting at their work."[57] Reporting on the racial situation in Pueblo, Dickerson wrote, "In general the employee at the steel plants is the common unskilled laborer and drawn from the poorest, most ignorant and illiterate class of Mexicans. Their appearance and demeanor is generally unattractive, if not worse, with the result that Americans and even other nationalities do not care to mingle with them and come to have a marked feeling of dislike and even contempt for them."[58] With such attitudes present in the city, Pueblo became a stronghold of the Ku Klux Klan during the 1920s. When one aggrieved steelworker, A. T. Jones, asked some other workers why they were laid off in 1927, they believed their failure to join the Ku Klux Klan was the reason.[59]

The easiest place to witness racism at CF&I's Pueblo properties was the Steel Works "Y." Despite the opportunities this $500,000 building offered, few steelworkers and their families took advantage of the wonderful facilities. "The membership [of the YMCA] has been very low, considering the total employment at the steelworks and the size of the community," explained the Curtis, Fosdick, and Belknap report in mid-1924.[60] The local "Y" director told Elton Mayo the reason for the lack of response. "Our challenge for service and activity comes from our Slavic, Italian, Mexican and poorer white employees residing within a mile or a mile and a half from our building," he said. "To develop activities which will bring these people into more expressive social and community relationships will require personal contact with them both in their homes, their neighborhoods, and with the men and boys in the mill."[61] This challenge was never met because Mexican and Mexican American workers did not feel comfortable in the building. "Generally speaking," Dickerson explained, "Mexicans do not now use the building and are to be found but rarely in the physical education or social features of the Association's program."[62] Since the YMCA was the focal point of much of CF&I's welfare work in Pueblo, the failure of non–native-born Americans to use the building significantly limited the effects of the company's welfare capitalist program.

EMPLOYEE REPRESENTATION AND CIVIL RIGHTS

William Dow was an African American miner who served as an employee representative at CF&I's mine in Rouse, Colorado. The minutes of the Joint Conference for the Walsenburg District, held in the nearby town of Ideal on May 16, 1919, report that Dow told those assembled that "he was very proud to have the opportunity of speaking for the men he represented; that the men at Rouse camp hold meetings often to discuss matters of interest, and those meetings are like love feasts, and peace and harmony prevails in their midst."[63] Yet Dow had been responsible for a serious grievance that had gone all the way up the chain of appeal to the president's industrial representative just a month earlier. Dow's complaint concerned segregation at the local YMCA. "I have been here for twelve years in this camp," he told the local industrial council on April 2:

> I think this [segregation] is a terrible injustice to us. The Industrial Plan gives every man a legal right. They have said if you wish to come to these entertainments you sit over here. . . . I feel that if the moving picture show is here and we come in and conduct ourselves properly, with our wives and daughters and we are dressed properly, I do not feel that it is right for the people here to rise up and say "you must sit over there."[64]

When CF&I management built a new clubhouse at Rouse just two years earlier, it turned over the old building "to the colored residents of Rouse and Lester" who "operated a Y.M.C.A. for their own race."[65]

Dow, in effect, wanted them to rescind that decision. The local YMCA council, unwilling to become involved in a racial quagmire, voted to leave the final decision on the matter to CF&I vice president B. J. Matteson. Shortly after the local industrial council heard Dow's complaint, management came to an understanding with the African American and white miners at the camp: "[T]he rule [about the separate YMCA for African Americans] previously passed by the white Y.M.C.A. council would not be enforced, but . . . the negroes would voluntarily sit together when attending public entertainments in the club building."[66] This disappointing result probably explains why despite praising the plan in May, Dow transferred from Rouse to Lester in December 1919, giving up his position as an employee representative in the process.[67] Certainly, this example suggests that the Rockefeller Plan did not function as well for Dow as its creators had intended.

Dow was not the only African American miner to serve as an employee representative. He was also not the only African American who used the rhetoric of the ERP to forward his civil-rights agenda. As Matteson explained in a report to President Jesse Welborn regarding a grievance from a group of black miners:

> I talked to several of the colored employees, and each of them showed me Mr. John D. Rockefeller's article on "Brotherhood of Men and Nations," as well as referring to the Industrial Plan; they stated that they had worked for the company at times when it was necessary for them to go armed in order to protect themselves and company property, and that some of the white fellows on this council were among the men on strike, and tried not only to destroy company property, but [also to] shoot men at work; and that the colored people, as a body, were not going to be subject to the dictation of that class of people.[68]

These employees might also have cited Rockefeller's article "Labor and Capital—Partners," in which he wrote, "[E]mployers as well as workers are more and more appreciating the human equation, and realizing that mutual respect and fairness produce larger and better results than suspicion and selfishness."[69] From a shop floor perspective, any "human equation" that made "mutual respect and fairness" possible had to include not just the relationship between employer and employees but also the relationship between employees of different races and ethnicities. When management told workers who had little power in the company because of their race that they were equal to their peers, it should come as no surprise that those workers tried to use the plan to forward their civil rights.

Ironically, even though management had been tracking the race and ethnicity of CF&I employees since at least the start of the twentieth century, John D. Rockefeller Jr. and Mackenzie King had barely considered the possibility that racial divisions could interfere with employees recognizing that aligning with their employer was in the mutual interest of labor and management. Such divisions, which practically define the study of modern labor history, were merely an afterthought to an industrialist trying to promote labor peace for the sake of production. In the past, CF&I had intentionally pitted miners of different races and ethnicities against one another in an effort to hinder worker solidarity. The Reverend Henry A. Atkinson described this policy in his report on the 1913–1914 strike:

> After the strike in 1883–84 the mines were operated with imported strike breakers, the Anglo-Saxon miners being in the minority. The newcomers were non-English speaking foreigners, men much inferior to the strikers whose places they filled. The strikers ten years later were these strike breakers who had been imported into the state ten years before. The strike was again won by bringing in another group of strike breakers, all foreigners, and viewed as laborers, an appreciably inferior class of men. Ten years later these men went on strike and, after deporting their leaders, the companies brought in men to take their places, men from southern Italy, Greeks, Slavs, Mexicans, Japanese, twenty-six nationalities in all.... It is these men who have found conditions intolerable and have struck.... The companies are fighting this battle in the same old way.[70]

The employee representation plan was a tacit denouncement of this strategy. Under this new system, management wanted employees to discuss their grievances in a public forum. The plan could not fix grievances that were racial rather than economic in nature, however, for fear of stirring tensions that could destroy the company from within.

Reflecting their lack of power within the company, Mexican and Mexican American miners generally did not make use of the plan to forward their position as a race. In contrast, other minority groups frequently used the Rockefeller Plan to advance their interests, much as William Dow did. For example, President's Industrial Representative David

Griffiths reported in June 1916, "Spent several days at Gulch adjusting a difficulty with nineteen Greek employees who had quit work as a protest because two of their number had been transferred from their regular working place. The Greeks finally returned to work without having their demand granted, and it was agreed that this was a case of misunderstanding." While Griffiths chalked this up as a victory for the plan, he also informed management: "They did insist, however, that Representative Marco Peters be transferred to some other mine. Peters is a native of the Island of Crete and the others are Grecians, and there is a little feud between the natives of the two locations."[71] In other words, one protest by an ethnic and numerically stronger group managed to overturn the will of the majority who had elected Peters in the first place. A note from the weeks before the 1919 strike described a similar ethnic-centered protest at CF&I's Sunrise Mine in Wyoming. "Two or three Greeks spoke about the houses of foreigners not being repaired," wrote the anonymous management author. "They seem to think there was some discrimination in that the foreigners' houses were not taken care of. One Italian spoke in behalf of other Italians."[72] Imagine the possibility that scores of these small, ethnically motivated, highly personal victories occurred throughout the mines and the mill, and it becomes easy to see why some CF&I employees who found themselves on the winning side of such disputes thought very highly of the Rockefeller Plan and some who found themselves on the losing side did not.

In contrast to the Rockefeller Plan, the United Mine Workers of America (UMWA) offered real hope as an interracial organization. Unlike many other unions at the time, the mine workers had been opposed to racial discrimination from the organization's inception. The first UMWA constitution in 1890 included a clause that read, "No member in good standing who holds a dues or transfer card shall be debarred or hindered from obtaining work on account of race, creed or nationality."[73] As historian Thomas Andrews has explained, the UMWA "probably made greater strides toward interethnic and interracial solidarity than any other major union prior to the New Deal."[74] While hardly a perfect democracy under President John L. Lewis, the union nevertheless could certainly compete with the Rockefeller Plan in its appeal to racial minorities.

When western mining concerns like CF&I began to hire Mexican workers, the UMWA reached out to them by hiring Spanish-speaking interpreters.[75] The results of these efforts were clear as early as the 1913–1914 strike. The legendary journalist John Reed reported on Mexican and Mexican American miners at the time:

> To them the union was the first promise of happiness and of freedom to live their own lives. It told them that if they would combine and stand together they could force the Boss to pay them enough to live on, and to make it safe for them to work. And in the union they discovered all at once thousands of fellow-workers who had been through the fight themselves, and were now ready to help. This flood of human sympathy was an absolutely new thing to the Colorado strikers. As one Mexican said to me: "We go out despairing and there comes a river of friendship from our brothers that we never knew!"[76]

Despite its aspirations to create a color-blind organization, the UMWA still had many flaws with respect to the issues of both race and democracy. However, CF&I miners, particularly Mexicans and Mexican Americans, would not have understood or perhaps even known that the independent union they sought to join had serious flaws.

John D. Rockefeller Jr. did not intend to have the employee representation plan that bore his name serve as a means of fighting racism and ethnic prejudice. Reflecting this hostility toward racial minorities, Rockefeller contributed at least $100 million for eugenics research even before the Ludlow Massacre occurred.[77] Eugenics was a movement inspired by Darwinian evolution that championed despicable policies such as selective breeding and forced sterilization to prevent inferior races of human beings from mixing with the gene pool of native-born Americans. Rockefeller's support for eugenics does not necessarily make him a racist. Considering his paternalism, Rockefeller's interest in eugenics probably sparked munificence rather than hatred. Nevertheless, a belief in eugenics had to undercut whatever support he might have had for racial equality.

In theory, Mackenzie King designed the Rockefeller Plan to channel class-based hostilities into discussions in which worker grievances could be resolved. Racial and ethnic hostilities, however, are almost by definition irresolvable through discussion because people cannot change the color of their skin or their ethnicities. Furthermore, Rockefeller's fondness for social order explains why the ERP Rockefeller helped design allowed skilled white workers, especially in the steel mill, to keep their less-skilled colleagues in subordinate positions. Intentionally or not, the Rockefeller Plan served as a means of social control for management to keep workers in line who belonged to races some whites deemed inferior.

CF&I executives nonetheless made grand claims that their union could serve as a vehicle for promoting equality between the races despite the financial risks management faced by doing so. For example, Welborn specifically invoked racial rights when comparing the plan to "the constitutions of the different States or the United States, which provide a basis whereby individuals, irrespective of race, creed, color or association, are entitled to representation in matters pertaining to their rights as citizens."[78] Similarly, management's growing acceptance of Mexicans and Mexican Americans as Americans can be traced in the pages of the *Industrial Bulletin*, culminating with the patriotic fervor of the World War I years.[79] However, the *Industrial Bulletin* represented only management's voice, not the voice of the white majority of employees. Even if a few nonwhite employee representatives were in fact equals, few of the company's minority workers had reason to think of themselves as such. As long as racists (or at least self-interested whites whose workplace objectives had negative effects on their nonwhite colleagues) dominated the ERP, management's promises of equality made no difference to racial minorities.

INDUSTRIAL RELATIONS IN THE MINING AND STEEL INDUSTRIES

Where racial and cultural differences divided CF&I's workforce, independent unions like the UMWA had to bring those workers together so it could organize them. In the mines, management created an employee representation plan so it could bring its employees together under its own authority rather than under the authority of the rival UMWA, which seemed to be waiting in the wings during the early years of the plan. An examination of ERPs from various countries from that period to the present reveals that companies often created these organizations only under the threat of organization by outside trade unions.[80] Nevertheless, when management introduced the Rockefeller Plan in the Minnequa Works in 1916, no such threat existed. This should be interpreted as another sign of John D. Rockefeller Jr.'s interest in promoting labor peace. However, management, particularly at the level of superintendents and foremen, implemented the plan differently in the two sides of the business because of the different contexts of industrial relations in the respective industries. This in turn affected how workers responded to the opportunities employee representation offered them.

Mining had a long history of trade unionism by the time CF&I created the Rockefeller Plan. The United Mine Workers of America had enjoyed great success representing midwestern miners (although it was much less successful elsewhere), and the Western Federation of Miners—formed by hard-rock miners in the Montana-Idaho region in 1893—had tried to organize Colorado miners repeatedly by 1915. Deeply involved in a series of eastern strikes, the UMWA did not put significant effort into organizing Colorado coal miners until 1907. The union was the driving force behind Colorado's Great Coalfield War in 1913–1914. Although never strong, the UMWA maintained a constant presence in the state throughout the history of the Rockefeller Plan. As stated in Chapter 3, many CF&I employees were active UMWA members. The mere availability of an independent alternative continually made the Rockefeller Plan look bad by comparison.

In contrast, the steel industry's lone independent trade union, the Amalgamated Association of Iron, Steel and Tin Plate Workers, had no presence in Colorado or most other states. It had been widely acknowledged as the strongest union in the United States until the infamous Homestead Lockout in 1892. While the union represented most ironworkers across the country, it never gained a foothold in the emerging Bessemer (or, later, the Open Hearth) steel industry. After Homestead, Carnegie Steel—the largest steel company in the United States—became union-free. When the Morgan interests bought Carnegie Steel and formed the giant U.S. Steel Corporation in 1901, the Amalgamated Association struck to leverage its presence in some plants and to gain a presence in others. It lost. By 1909, U.S. Steel, which produced approximately 65 percent of the product in the entire industry, was union-free, and the Amalgamated existed in only a few midwestern mills that made specialty steels in a manner that still required the union's skilled members. The 1919 steel strike nearly destroyed the Amalgamated Association of Iron, Steel and Tin Plate Workers. Membership in the union dropped from 31,500 in 1920 to only 4,700 in 1933. Every labor dispute the union became involved in during this period resulted in a loss of a previously organized mill, and it organized no new mills over this same time frame.[81]

From 1916 through 1933, CF&I miners showed every inclination to organize while the company's steelworkers only struck once, in 1919 (and, as will be explained in Chapter 6, that strike occurred almost by accident). Nevertheless, at least in the early years, upper management still treated the implementation of the Rockefeller Plan in both sides of the business with great care. After all, employee representation was not just a tactic; it was a philosophy for John D. Rockefeller Jr. His reputation was at stake. Lower management (all the way down to foremen) had no such compunction. Management replaced superintendents in the mining division with men who had "more modern ideas" and who helped implement the plan smoothly.[82] Recognizing that little or no outside union threat existed, foremen and superintendents at the Minnequa Works often saw no good reason to make the kinds of concessions management in the mines regularly offered. In return, the employee representatives there became more combative than those in the mines.

Miners could not help but compare the activities of the ERP with the actions of the United Mine Workers of America. Since the union served as a reminder to those workers of what an independent union could accomplish, management did its best to match or even exceed union benefits so its workers would have no need to organize. Steelworkers, on the other hand, had no union to use as a basis for comparison. They judged the plan on its merits and found it lacking. This may seem odd in light of the fact that the Rockefeller Plan lasted almost a decade longer in the steel mill than it did in the mines, but an overwhelming number of contemporary sources support this argument. In 1927, for example, investigators from Curtis, Fosdick, and Belknap concluded, "[T]he joint representation conference plan met with a more intimate and prideful endorsement by the employes, supervisors and officials of the fuel department than it did by the personnel at [the] Minnequa Works."[83] After his 1928 visit Elton Mayo reported, "[A]t the mines there seemed to be a feeling of greater confidence on the part of employees and management, causing the Industrial Plan to function better than it has at the Steel Works."[84] So why, then, did the Rockefeller Plan end sooner in the segment of CF&I in which it operated more effectively? The answer involves the inherent limitations of any employee representation arrangement.

In 1921, Harvard-graduate-turned-itinerant-socialist-worker Powers Hapgood wrote about his experiences at two mining camps for *The Nation*. One was CF&I's Frederick Mine in Valdez, Colorado. The other was "Brophy's" Mine at Bearcreek, Montana. "The living conditions at the Frederick Mine . . . are excellent," he wrote:

> Near the mine there is a well-kept main street lined with attractive houses in which the married miners live in comfort with their families for very low rents. A white school house is located just beyond the houses, and a good-looking brick building stands out conspicuously as the center of the village life. This is the Y.M.C.A. built and maintained by the company for the benefit of its employees. Here there are moving picture shows twice a week, and now and then a dance is held in the large reading room. The bowling alleys and the pool tables in the basement and the checker boards and other games upstairs afford opportunity for light amusement for the miners, while a reading room with books and magazines gives a chance for the more serious minded to educate themselves. Evening classes are held for those who wish to advance from their places as coal diggers to positions as fire bosses, mine foremen and mine superintendents. Men who are injured at work are taken care of by the company in return for a small fee each month from the individual miners, and in the case of death of an employee of the company his family is permitted to remain in the house which it occupies as long as it wishes free of charge. A weekly income is paid to the widow of a miner who is killed, with an additional amount for each child.[85]

In contrast, at the Bearcreek Mine there were

> scarcely any of the benefits enjoyed in the mining camps of the Colorado Fuel and Iron Company. Most of the houses might well be called shacks, and, while the little coal town of Bearcreek, surrounding which there are seven or eight mines, boasts a moving picture show and what used to be called saloons, there is no Y.M.C.A. or place where the miners can read and attend classes of instruction. The scale of wages is a little higher, but this in no way makes up for the lack of Y.M.C.A.s, well-built houses, and benefits to families of deceased miners.[86]

Someone unfamiliar with the situation at CF&I might think its workers were represented by the UMWA and those at Bearcreek were not, but in fact it was the other way around.[87] Perhaps even more surprising, the miners at Bearcreek were much happier than those who worked for CF&I. According to Hapgood, the miners at Bearcreek "were conscious that they were independent as a group and that each of them had a voice in his own destiny, and because of this they had gone a long way toward acquiring that feeling of self-respect which is so necessary to the happiness of normal men."[88] Ben Selekman made a similar observation when comparing CF&I with another briefly unionized Colorado mining firm. "[G]reater spiritual freedom seemed to prevail in the Victor American camps than in the C.F.&I. camps," he wrote, "in spite of the fact that physically the C.F.&I. camps appeared as prosperous city suburbs as compared with the miserable conditions prevailing in the Victor American camps."[89]

Mackenzie King and John D. Rockefeller Jr. had not anticipated this reaction. They assumed that welfare capitalism and limited rights would translate into widespread support for the ERP. However, independent unions offered intangible benefits that even the most liberal company-sponsored employee representation plan could not match. A consultant on another industrial relations matter wrote to Rockefeller about this dynamic with respect to another mine in the late 1920s:

> There is a psychological appeal in labor unionism which has not yet been analyzed. It seems to give the men a sense not only of power but of dignity and self respect. They feel that only through labor unions can they deal with employers on an equal plane. They seem to regard the representation plan as a sort of counterfeit, largely, perhaps, because the machinery of such plans is too often managed by the employers. They want something which they themselves have created and not something which is handed down to them by those who pay their wages.[90]

Despite its careful design, the Rockefeller Plan could never address the vital element of independence outside trade unions provided. The clearest way to see how both the benefits and limitations of the Rockefeller Plan played out on the shop floor is to look at its operation in detail in both sides of the business, as explored in Chapters 5 and 6.

Chapter Five

THE ROCKEFELLER PLAN IN ACTION: THE MINES

> The miners never lost anything at any of them [Rockefeller Plan] meetings that I ever know of. . . . They got everything they asked for at these meetings.
>
> —FORMER CF&I MINER BILL LLOYD, MAY 18, 1978[1]

In the late summer and fall of 1919, before Colorado Fuel and Iron (CF&I) became embroiled in a nationwide coal strike, Paul F. Brissenden of the U.S. Department of Labor conducted a series of meetings across southern Colorado among union miners, non-union miners, and local managers. On August 13 the superintendent of the company's Starkville Mine wrote Elmer Weitzel, the manager of all CF&I mines at the time, about one such meeting there. "They took a standing vote on the Rockefeller Plan and everyone voted against the Plan but three men," he reported.[2] On August 15 Weitzel wrote CF&I president Jesse Welborn about a similar meeting at Sopris with 130 miners. "When the vote on the Industrial Plan was taken," he reported, "the vote was unanimous against the plan." At Pictou the miners held a secret ballot referendum on the Rockefeller Plan, with fifty-nine voting against the plan and only four voting for it.[3] As Brissenden later explained, "At most of the meetings the great majority of those in attendance were members of the Union, although there was in nearly every case a group of non-union men present. I recall talking personally at these meetings with a number of non-union men. . . .

In some of the mines the great majority of the employees are members of the United Mine Workers."⁴ Welborn later received detailed reports of all the meetings from Brissenden.⁵ Had management actually listened to the reports of the meetings, they would have recognized that the Rockefeller Plan was not nearly as popular as John D. Rockefeller Jr. and Mackenzie King had hoped it would be.

Instead of seeing the feedback from Brissenden's meetings as one of many signs that the Rockefeller Plan had serious problems, CF&I attacked every independent messenger who tried to alert the company to problems with the plan. In a letter to Congressman Claude Hudspeth, the company accused Brissenden of "deliberately using the power of his office, and the good feeling towards a federal officer, in the mind of a private citizen, to aid and abet the United Mine Workers' organization and organizers to accomplish through the prestige which his office afforded a dissatisfaction which a persistent effort of three years by these same men had failed to accomplish," even though CF&I had known that a great deal of discontent existed before Brissenden ever arrived on the scene.⁶ Similarly, in September of that year, when nearly identical resolutions signed by miners at ten locations arrived on Welborn's desk—foreshadowing union demands during the strike later that year—the company's president continued to claim that the Rockefeller Plan had support from a majority of coal miners. "Therefore," he wrote in a form letter to each set of disgruntled miners, "your communication cannot be regarded as expressing dissatisfaction on the part of our employees with the Representation Plan."⁷

Scholarly researchers received similar treatment. In December 1920, Mary Van Kleeck of the Russell Sage Foundation presented CF&I with two long manuscripts written by Ben M. Selekman describing the operation of the plan in both the mines and the steel mill. Selekman based these studies largely on candid interviews with employee representatives and employees. While Van Kleeck expected that the company would appreciate these new insights into its industrial relations policy, it preferred to challenge the messenger. "We are all very much disheartened by it," Rockefeller wrote to Mackenzie King after receiving the reports, "for although the officers of the [Russell Sage] Foundation have purposed to make a perfectly unbiassed [sic] report, we cannot help but feel that the investigator has been strongly biassed [sic] against the plan, and that his report, unless very materially modified, will discourage the weakening interest of other employers in some form of representation."[8] Managers from John D. Rockefeller Jr. on down judged the success of the employee representation plan (ERP) on the basis of the lack of grievances they received, and Welborn seldom heard complaints from his miners. As the fire boss at the Cameron Mine wrote Rockefeller with respect to Welborn, "I have seen him at a gathering of the officials and miners, literally take his coat off and work for hours trying to get them to come into the open with their grievances; but the local people distrust one another and altogether distrust the General Office."[9] Perhaps Welborn assumed that if the representatives did not make their grievances known to him, there were no grievances to voice, never realizing that the miners feared expressing their true feelings to the president of the company. Even when workers complained directly to management under the auspices of the plan, company executives took this as proof for the fiction that the two parties were working out their differences to their mutual satisfaction.

This chapter and Chapter 6, which discuss the operation of the ERP in both sides of CF&I's business, sample many of the minutes of plan meetings available in the CF&I Archives in Pueblo and other locations. The chapters demonstrate that many of the miners and steelworkers who served as employee representatives took the promise of the ERP to heart, at least to the extent that they willingly (and sometimes angrily) expressed their grievances and those of their constituents to management at regular meetings. However, even though some employee representatives tried to use the plan to improve conditions at the company, not every worker necessarily took it seriously. Available records include much discussion of how to get more employees interested in the day-to-day workings of the ERP. The employee representatives grew increasingly frustrated as time passed and they came to realize the limitations of operating under a system dominated by their employer.

"THERE ARE DAYS WHEN I GET NOTHING ELSE DONE"

Colorado coal miners in this era, explains Thomas Andrews, worked at their own pace. Most were paid by the ton, worked with their own tools, and were rarely visited by their supervisors, so the workers had great control over their effort and actions while on the job. Furthermore, shared

THE ROCKEFELLER PLAN IN ACTION: THE MINES

safety concerns formed the basis of solidarity that served miners well when they chose to undertake collective action aboveground.[10] The Rockefeller Plan was supposed to undermine this oppositional culture and persuade miners to align their interests with management. Miners used the ERP to air complaints big and small. Some of these complaints could be resolved, but many others could not because labor's interests and management's goals in the workplace were fundamentally contradictory. Miners wanted to get more for less effort, and their employer wanted the opposite.

Contemporary critics of the Rockefeller Plan tended to see the plan's potential harm as being that it created the impression that it would actually satisfy demands workers placed on management. As George P. West of the U.S. Industrial Relations Commission wrote in that body's report on the 1913–1914 Colorado coal strike, "The effectiveness of such a plan depends wholly upon its tendency to deceive the public and lull criticism, while permitting the company to maintain its absolute power."[11] But the company did not maintain absolute power. Records of the meetings of labor and management conducted under the auspices of the Rockefeller Plan show that many employee representatives forcefully voiced grievances to lower-level management and that management frequently gave in to those demands. In fact, many of the meeting minutes read more like gripe sessions than dialogues. Workers sometimes praised management as well, and the same miners could be either belligerent or sycophantic, depending on the circumstances they faced. While miners would have had more freedom of action had they been represented by the United Mine Workers of America, they still managed to turn the Rockefeller Plan into a successful bargaining vehicle.

Critics of the ERP, beginning with Selekman and Van Kleeck in the mid-1920s, have seized on the relatively low number of complaints from throughout the company that reached the president's industrial representative as the basis of their disapproval of the plan. In 1920, management reported that only 85 of 5,556 people on the entire company payroll filed complaints.[12] Certainly, some CF&I employees did not feel comfortable airing their grievances through the avenues provided by the ERP. "When [David] Griffiths [the president's industrial representative] came here," a former representative told Selekman and Van Kleeck, "I have to tell him everything is all right. If I go to the men, they say they have grievances, but when I ask them to come and prove it to the superintendent or Griffiths, they say they are afraid. So I tell Griffiths everything is all right."[13] There were thus two different stages at which grievances could have died: in the minds of the men and in the hands of employee representatives. With abundant opportunities for problems, it should be considered miraculous that so many grievances actually did reach managers. With the plan in regular use, management probably saw no need to dig deeper to find out how employees really felt.

Nevertheless, the figures Selekman and Van Kleeck used to argue that the plan was underutilized are highly misleading because they do not include informal complaints that did not reach the formal hearing stage. It is impossible to quantify the exact number of complaints a foreman or superintendent handled over a particular period because the company did not keep track of every grievance. Throughout the history of the ERP, however, management solved many grievances informally before the president's industrial representative had to be summoned. Even when that representative did become involved, the grievances he handled were not always written up for management in Denver.[14] The number of informal grievances appears to have been staggering. "There are days when I get nothing else done, as my time has been practically all spent in listening and advising and suggesting," wrote C. A. Kaiser, superintendent of the Berwind Mine, in response to a circular letter requesting feedback from foremen on the results of the plan.[15]

This situation suggests a paradox. How can workers be intimidated by the plan and not intimidated by it at the same time? It depends upon which workers one considers. Rank-and-file miners, especially Mexicans and Mexican Americans, made little use of the plan, as discussed in detail in Chapter 4. As Representative William Gilbert explained during a plan meeting in 1917, "There are times when matters come up that cannot be settled through their representative."[16] Therefore, more skilled and longer-term workers accounted for most of the informal grievances. The success of higher-skilled white workers at winning grievances likely contributed to that group's longevity in the company's employ, especially in the steel mill. This is reflected in the nature of the issues that came before the ERP. For example, after analyzing the figures from 1922, President's Industrial Representative B. J. Matteson found "in many instances the question was merely a controversy between two employees to find out which one was entitled to a certain job and that it was immaterial to

THE ROCKEFELLER PLAN IN ACTION: THE MINES

5.1. *Employee representatives, managers, and their sons from Berwind at a dinner in 1923. Courtesy, Bessemer Historical Society.*

the Company which one held the job as long as the operations were not interfered with and harmony was maintained."[17] To have a claim to a certain job, one needed to have the kind of tenure Mexicans and Mexican Americans seldom enjoyed.

During the first six months of 1922, Matteson estimated that two complaints were filed somewhere in the company each day. One of the two resulted in an adjustment (meaning that the employee initiated discussion that led to some kind of change in his favor).[18] This is a record any independent union would envy. Nevertheless, the value of these employee victories must be regarded with skepticism. It seems likely that most of the worker victories involved matters of little importance to management. As the superintendent of the Berwind Mine told John D. Rockefeller Jr. during his first visit there, "[T]he men were constantly coming to the representatives with trivial matters which ought never to come up."[19] Thus, to gain an idea of whether the Rockefeller Plan truly satisfied employee representatives and their constituents, it is necessary to examine exactly what employee representatives did while fulfilling the duties of their position and how they used the power the plan gave them. This chapter will consider the operation of the plan in the mines; Chapter 6 discusses its operation in the steel mill.

THE WORK OF THE JOINT COMMITTEES

Most of the day-to-day disputes considered under the auspices of the plan in the mines started in the joint committees. As briefly explained in Chapter 3, there were four joint committees (Industrial Cooperation and Conciliation; Safety and Accidents; Sanitation, Health and Housing; and Education and Recreation) in each of the five mining districts (six once the steelworks came under the ERP). The committees brought the results of their deliberations before the district joint conference after hashing out issues and drawing up proposals. The chair of each committee was always a member of management. Miners from different camps served on each joint committee, and the committees could hold meetings in any camp they chose. Representatives elected three of their own members at the first meeting of each year. The *CF&I Industrial Bulletin* published reports on the activities of the joint committees in special issues throughout the life of the plan.

The first subcommittee mentioned in the original plan was the Joint Committee on Industrial Cooperation and Conciliation (later the Joint Committee on Industrial Cooperation, Conciliation and Wages).[20] The purpose of this committee was to work toward "the prevention and settlement of disputes, terms and conditions of employment, including wages, hours and working conditions; maintenance of order and discipline, company stores and other similar matters."[21] (As the debates on wages throughout the company went beyond the committee's activities, this subject merits separate analysis, presented later in this chapter.) Like the employee representation plan itself, each committee consisted of an equal number of representatives from labor and management. Testifying before the Colorado Industrial Commission in 1928, B. J. Matteson claimed that none of the committees had ever deadlocked in the twelve years since the plan had begun.[22] This strongly suggests that these bodies had the same flaw as the one inherent in the plan itself. Management maintained the upper hand when deciding controversial issues, and labor knew it. Even if workers wanted to appeal a grievance all the way to the

Colorado Industrial Commission, they knew managers could make their lives extremely difficult in the interim. Yet despite this flaw, the committees' activities improved at least some conditions for workers at both the mines and the mill.

The Joint Committee on Safety and Accidents dealt with perhaps the most important issue labor and management faced under the auspices of the Rockefeller Plan. Unlike wages, which were initially out of the employee representatives' hands, this was literally a life-or-death issue, and better communication should have made improvements in safety easier to achieve. CF&I's mines desperately needed better safety rules and enforcement. In the years before the 1913–1914 strike, the company had been rocked by a series of fatal mine explosions—19 dead at Tercio in 1904, 19 killed at Cuatro in 1906, 24 dead at Primero in 1907, and 75 killed also at Primero in 1910.[23] Even in 1911, a year with no massive mine explosions on CF&I property, 22 company miners died from fatal accidents.[24] As a result, miners made safety a major issue during the 1913–1914 strike.[25]

There is no question that working in CF&I mines became safer after the plan took effect. "[B]y the mid-1920s," James Whiteside has argued in his book about safety practices in the Rocky Mountain coal mining industry, "CF&I's safety program included more attention to timbering and shot firing, more sprinkling in rooms and entries, electric illumination in entries, attention to safe installation of electric power and trolley lines, and safety guards on machinery." However, CF&I had begun to make significant safety improvements before the Rockefeller Plan took effect. For example, it put the first mine-rescue car in the nation into operation in 1910.[26] In 1915, without consulting its recently constituted ERP safety committees, management instituted a new system of incentives and direct supervision over mine safety.[27] For such reasons, it is impossible to directly attribute specific safety improvements at CF&I to activities undertaken under the auspices of the plan.

Furthermore, the manner in which the Joint Committee on Safety and Accidents operated in the mines became a significant source of tension between labor and management. Testifying before the Colorado Industrial Commission in 1928, a lawyer representing striking miners asked the head of CF&I's mining division, R. L. Hair, "In how many cases did the committee find the accident occurred by some failure on the part of the Company or its management?" Hair replied, "Well, I do not know of any time they came out and directly stated it was directly caused by the Company as directly the cause of the person being injured or the person killed."[28] This suggests that management used the committee to cover up its own negligence. The records of the Colorado Industrial Commission offer countless examples of injured workers or dead workers' families finding redress in a less partial venue. For example, Cesario Mondragon died when he touched a live wire that ran the man trip that carried miners out of his mine. Management claimed it should not pay compensation because Mondragon broke a safety rule by boarding the man trip at the wrong place. The commission rejected management's request, explaining "[t]hat the evidence submitted indicates that the employes of the employer repeatedly and continuously violated said alleged rule and with full knowledge of said violation on the part of the employer."[29] This explains why miners seldom took safety rules seriously. Their implementation often constituted just another way for management to wield unchecked power. Such examples demonstrate a failure of communication between labor and management that the framers of the ERP had not expected.

Despite its professed desire for safety, the implementation of the plan did not change a dangerous system that encouraged both miners and mine foremen to downplay safety in order to increase production.[30] For example, seventeen miners died in an explosion at Sopris in 1922. "Just what caused the explosion may never be ascertained definitely," reported *The Trinidad Evening Picketwire*. "The mine had the reputation of being a safe one and was equipped with every preventative device to minimize dust and gas explosions."[31] This indicates that the company did not understand exactly how to make its mines safer. Members of the Joint Committee on Safety and Accidents supposedly accompanied CF&I's mine inspector on his tours of their facilities and made recommendations for improvements to superintendents, even though company officials ignored safety rules themselves.[32] Yet in 1927, the fatal accident rate for miners in southern Colorado (which CF&I dominated) was 57.8 percent higher than the rate for miners working in northern Colorado (where CF&I had no mines).[33]

Despite CF&I's seemingly abysmal safety record, the company actually ranked somewhere in the middle on safety in the context of the total Colorado coal industry throughout the Rockefeller Plan era. In 1917 the state mine inspector declared, "[W]e have had less trouble [in CF&I's mines], and have had our recommendations carried out promptly." This "would leave the inference that [CF&I's] mines are in better condition than other mines."[34] That situation no doubt resulted in part from the existence of the safety committees. Of Colorado coal mines that had safety committees, the others were in mines with strong union backgrounds.[35] "Lots of times," argued Frank Hayes of the United Mine Workers of America (UMWA), "[a miner would] rather take a chance with his life than take a chance with his job." The Rockefeller Plan allowed CF&I to gain some, but not all, of the benefits of an independent union with respect to safety matters. The presence of a truly independent union would have done more to rectify this problem by offering a whistle-blowing miner job security even if he criticized his superiors' safety practices.[36] Indeed, organized mines had 40 percent fewer fatalities than did non-union mines during this era.[37]

The Joint Committee on Sanitation, Health and Housing also had enormous potential to improve conditions in the mining camps. Selekman and Van Kleeck were very complimentary, for example, of improvements CF&I made to its workers' housing stock. "[A]lthough many of the houses had been built which now appear so attractive, and the spruce trees native to the place had been permitted to remain, the care of the housing was not provided for as it was after 1915," they wrote. "In earlier days there had been no fences, no yards, and therefore no scope for the pride of each miner's family in the care of its own plot of ground."[38] Even George Johnson, president of the local UMWA, agreed with this assessment. "He could hardly believe his eyes when he visited the C.F.&I. camps last year," explained Ben Selekman in a summary of his 1919 interview with the union leader. "There is no doubt but they have the best houses, keep their camps in the best condition and have the best recreational facilities [of any of the mining companies]."[39] Nevertheless, even miners who lived in these superior communities continually asked for more to be done at management's expense.[40] No matter how comfortable their existence compared with that of other industrial workers across the nation, rising expectations still made some employees dissatisfied with their living conditions.

The work of the Joint Committee on Education and Recreation included such popular activities as coordinating the annual company picnic (known as Field Day) and helping to run the YMCAs in the mining towns and at the steelworks in Pueblo. "Throughout the camps there seemed to be a real pride and interest in the Association as a community enterprise," reported Roy Dickerson of the YMCA during his 1921 tour. "There seems to be a general feeling that there has been a great change in the camps since the incoming of the Association due largely to its efforts and both men and superintendents are appreciative of the fact and proud of the Association." Some of that appreciation undoubtedly stemmed from the miners' ability to control their own social activities through the recreation committee, since the operation of the YMCA was actually under the control of CF&I and therefore subject to discussion under the auspices of the plan. Dickerson's report on Mexican industrial life in the coal camps noted that those workers needed "to be given an opportunity to have a part in the development of their own program. This means for one thing that Mexican representation on the councils is very desirable."[41] The non-Mexicans who controlled the day-to-day operation of the recreation centers, however, certainly benefited from this opportunity they received through the plan.

SUSPENSIONS AND DISMISSALS

While much of the work of the ERP was of little long-term importance to the direction of the company, many other grievances involved individual workers' livelihoods and were therefore very important to them. In 1920 management displayed a poster listing a series of offenses that, if committed, would lead to a miner's immediate dismissal. The offenses included "carrying a concealed weapon, fighting, bootlegging, offering or receiving money in exchange for a better working place, attempting to organize or spread propaganda of the Industrial Workers of the World or other radical organizations, and 'efforts to disturb harmonious relations within the company.'"[42] While this last category left open a large area for abuse, miners who wanted to keep their jobs with CF&I invariably turned to the ERP for redress and often won their cases.

Meeting minutes from the mines cite many arguments over dismissals and reinstatements of miners, including some who were reinstated even though they probably should have been permanently dismissed. For example, at the mine in Lester in December 1922:

> Three employees, who were discharged for stubbornly refusing to get out of one car and into another car being pulled into the 7th north entry, saying that if the car they were in did not go in that they would not, took question of reinstatement up with representatives, and after being reprimanded for their conduct they were told they could go back to work if the mine foreman could use them. They were given work digging coal, and later two of them got their old jobs back.[43]

A worker at Toller was reinstated in March 1922 after committing an offense with more serious consequences for production:

> Ellis Williams, machine runner, broke [a] chain on mining machine early one evening. Mechanic and electrician promptly responded to call for repairs, which were made within few minutes after reaching machine. Williams went home, instead of remaining to cut his place after repairs were made, thus causing loaders to be short coal, and was given his time for delaying operations. Williams complained through representatives . . . and was allowed to return to work.[44]

In 1918, an officer of the United Mine Workers of America used the Rockefeller Plan to get his job back at the Lester Mine after taking an unauthorized month off to attend the union's national convention in Indianapolis.[45] In 1919, miners at Engle staged a strike to have a laid-off employee reinstated. After employee representatives met with management, the man returned to work. The strike lasted one day and affected only that mine.[46] In 1924, the superintendent dismissed a long-time employee at Ideal for soliciting other miners to join the Industrial Workers of the World (IWW). The employee representatives filed a complaint; the president's industrial representative came from Denver and eventually had him reinstated. (Ironically, the miner joined the 1927 IWW strike, thereby demonstrating the truth behind the initial firing.)[47] Even though a few disciplinary actions were eventually upheld, it is understandable that the foremen at Lester would write in 1924 that some men "are inclined to believe that it is impossible to discharge them."[48]

Many discharge appeals likely succeeded because of the effect a worker losing an appeal had on other workers. Consider the firing of two miners, James Weir and Max Ratkovitch, at the Coal Creek Mine in April 1918, a case detailed in depth by Selekman and Van Kleeck. A superintendent told the two miners to stop "brushing" a wall in the mine. When the superintendent returned several days later and saw that they

THE ROCKEFELLER PLAN IN ACTION: THE MINES

had brushed the wall against his order, he fired them for insubordination (one of the offenses for which a miner could be fired without appeal). The two workers claimed they were obeying the order of their foreman, who backed them up and threatened to quit if Weir and Ratkovitch were discharged. President's Industrial Representative David Griffiths was convinced that the two men had been discharged unjustly. In fact, he argued that the superintendent had violated state mining law because foremen were supposed to have complete charge of underground operations. At first the superintendent agreed to take the men back, but he transferred them to a different part of the mine. Weir and Ratkovitch complained that the new location was not as good as the old one. As a solution, management transferred them both to CF&I's mine in Rockvale.[49]

Ben Selekman found many miners willing to express dissatisfaction with this result when he arrived in Coal Creek two years after the incident happened. Employee representatives and regular miners alike pointed to the fact that the superintendent had not been overruled—indeed, he was still superintendent of the same mine—while the innocent miners had to suffer through a transfer. More important, "They pointed to this case as illustrating a fundamental weakness with the plan. They thought that it meant company officials would sustain a superintendent against a miner. How could the miners have confidence in the plan, they asked, when the only protection they have under it is the fairness and impartiality of the decisions made by management?" Management had ignored the pleadings of the two employee representatives who took Weir and Ratkovitch's side.[50] Management might have been able to stem some of the anger over dismissals if it had set up a system by which a joint committee of management and employee representatives reviewed all dismissals as they took place, but such a system was never instituted. Instead, labor gradually chipped away at management's prerogatives on a case-by-case basis.

The best evidence for a decline in management's power over time as a result of the plan is the disappearance of these lines in the consolidated version of the ERP: "Nothing herein shall abridge the right of the Company to relieve employes from duty because of lack of work. Wherein relief from duty through lack of work becomes necessary, men with families shall, all things being equal, be given preference." The minutes of meetings of the committee that revised the plan have not been found, so it is impossible to know exactly why management forfeited this reserved right in the 1921 version. Was it simply related to the fact that the Minnequa Works did not resort to layoffs until the 1930s, or was it perhaps a reflection of management's overall inability to fire anybody? Either way, the absence of this clause in the revised ERP clearly speaks to a decline in the company's power vis-à-vis its employees.

WAGES

Two clauses in the signed memorandum that accompanied the initial ERP dealt with wages (see Appendix 1). The first stated: "No change affecting conditions of employment with respect to wages or hours shall be made without giving thirty days notice, as provided by statute."[51] The second clause read, "The schedule of wages and the working conditions now in force in the several districts shall continue without reduction, but if, prior to January 1, 1918, a general increase shall be granted in competitive districts in which the Company does not conduct operations, a proportional increase shall be made." Management designed these provisions to take wage issues off the table at ERP meetings, even though the second clause indirectly allowed the United Mine Workers of America (which had organized most eastern mines) to negotiate wage increases for CF&I miners. Yet, as has been described, some miners earned much more than others. The highest-paid skilled miners, as Elmer Weitzel explained to the U.S. Senate Committee on Interstate Commerce in 1920, could afford their own automobiles.[52] Other miners could barely make ends meet.

There were two reasons for this disparity. The first, as discussed in Chapter 4, was intermittent employment. No matter how high or low wages may have been, miners still had to work to earn anything. Even the most successful workers paid by the ton seldom had the opportunity to work more than twenty days per month in lean times (and most of the 1920s were lean times for CF&I mines). The second reason for the wage disparity was the piece rate system. Changing that system directly through the employee representative plan would have been such a radical innovation in the coal mining industry that management would have rejected the idea outright, but many grievances in the available ERP minutes report that employee representatives tried to make small inroads in that direction. For example, consider a 1917 request from employee representatives in the Canon District: "It was moved and seconded that the representatives ask for a raise of 5% in the place of the bonus now in effect. The cause of the request is that a man who is making good pay gets so much more than a man who has a poor place and cannot make so much, even if they work the same number of days." Management turned the proposal down flat because "the bonus was put into effect as an inducement to the men to work every day in order that the largest output possible might be had while the present good demand for coal existed."[53] Management gave bigger bonuses to workers who had a greater role in determining output because it was concerned more about production than about equality. Even though miners had no right under the plan to influence wages, they still tried to introduce justice into the system, just as independent trade unions often did.

Another key wage-related issue was the availability of equipment for men on the job. Minutes from the First Joint Conference in the Canon District from February 1922 reported:

> Representative Thomas Payne said he had been informed that many of the good miners in the Rockvale new slope mine were not able to load enough coal to make satisfactory wages, due in part to the large amount of brushing and shortage of cars. After a general discussion, all agreed that the effort to provide many more men with work than were necessary to take care of the operation, was principally responsible for the low average earnings of miners, but, under the circumstances, this situation was preferred to laying off men.[54]

In other words, management could have discharged people because of a lack of work but chose not to exercise that prerogative, probably as a result of input from employee representatives. With more men and fewer cars, it did not matter how fast a miner could dig coal; he was still unable to earn all he could have potentially earned because there was nowhere to hold his product.

After World War I ended, CF&I had problems when it tried to follow the wage trends that emanated from collective bargaining with the UMWA in the East. Workers (like any other economic actors) are reluctant to make concessions even if the need for those concessions is explained to them in detail. Nevertheless, in 1921 management tried to use the bargaining structure established by Rockefeller and King to communicate its position and thereby make it easier to implement a wage cut that started in the eastern fields. Before it took this step, management started laying off miners for lack of work. According to the Colorado Industrial Commission, between March and September of that year, "the number[s] of days the majority of the mines in this state have operated have been so few that the miners working therein have scarcely made a living wage."[55] This should have led miners to be willing to accept more work at lower wages. Instead, when management announced its long-rumored wage cut effective September 1, many southern Colorado miners (including some not even affected by the wage cuts) went out on strike.[56] Huerfano County, Colorado (home of several CF&I mines), declared martial law during the dispute.[57] However, the strike ended quickly after a group of employee representatives petitioned for a substantial wage decrease to get their mines working again (on the theory that work at a lower wage was better than no work at all).[58] This brought enough employees back to work to allow the mines to run again.

Essentially the same dispute occurred again in 1922. In this instance, management sent a delegation of steelworker-employee representatives to tour the coal camps, explaining why the miners should accept another across-the-board wage decrease under the auspices of the ERP. The motivation for steelworkers to have made the trip is easy to discern—without coal they could make no steel.[59] This time miners in Cameron refused to accept the pay cut and went on strike. A few days into the dispute, a group of miners again petitioned management to reopen the mine, and enough workers appeared to break the strike.[60] The existence of an arrangement that gave workers a voice could not prevent labor trouble in these instances. Indeed, it may only have reinforced the limitations miners faced without their own independent union.

THE UMWA AND THE ROCKEFELLER PLAN

In its story on Rockefeller's 1915 speech introducing his plan to miners in Pueblo, *The Pueblo Chieftain* reported that six of the men who ratified the ERP that day were members of the United Mine Workers of America.[61] *The Denver Post* put the figure at eight.[62] "The plan looks good and I can't see that it interferes in any way with the union," one of the union members told the *Chieftain*.[63] These comments reflect the UMWA's initial strategy in response to the Rockefeller Plan—utilize the ERP to improve conditions for its members, but continue to organize until it had enough members to demand that management sign a union contract instead. Two years after the plan was introduced, journalist John Fitch was surprised to see the president of the Colorado Federation of Labor and the acting president of the United Mine Workers of America for the district covering Colorado leaving a meeting with Jesse Welborn.[64] Fitch reported that the UMWA claimed that 90 percent of the men in the firm's mining camps were members. CF&I officials thought the percentage was closer to half.[65] Both figures would have been large enough numbers to create trouble. In fact, local union president George Johnson actually adjusted grievances between CF&I miners who belonged to his organization and local superintendents.[66] It must have seemed as though the company was on the verge of recognizing the union.

It did not do so, however. As explained in Chapter 2, Rockefeller believed in a fair competition between his union and the UMWA but did not believe his union would lose that contest. One of his tactics for winning the competition was to insist that management protect at least some of the rights of CF&I employees. In 1916, for example, Rockefeller wrote to Welborn suggesting that the company "issue an explicit notice to the employes, setting forth the rights that they enjoy and also the protection which the company will give to them in the exercise of their freedom of association or non-association with any organization." Management's letter in response to Rockefeller's suggestion read in part: "[E]mployes have a right to hear what any visitor has to say, and to determine each for himself what action he will take. The fact that any employe belongs to a union or other organization will not affect his interests with this company."[67] Even if the miners did not believe it, the company meant what it said, at least at this juncture.

In response to this expressed toleration of outside unions, the number of UMWA miners at CF&I swelled and union-related activity proliferated rapidly in the camps. Miners at both Berwind and Tabasco struck for a day in 1916. In August of that year, the houses of two employees at Tabasco were dynamited because of their support for the Rockefeller Plan. On February 5, 1917, miners at Ramey in Huerfano County struck for a 20 percent increase in wages; they received a 10 percent increase.[68] When the Colorado Industrial Commission reported its findings on an aborted company-wide strike in October 1917, it found that "[t]he Company employs many members of the Union, and when employing men does not inquire as to their membership in labor organizations."[69] Furthermore, as explained earlier, in 1919, the year of a nationwide coal strike, even the company admitted that half its miners were UMWA members. If management intended the Rockefeller Plan as a device to destroy the union, it failed miserably.

Yet despite the company's tacit acceptance of the union's presence in its camps, the UMWA still had good reason to be hostile to the ERP. Even though management allowed union meetings in its camps, it also intended the employee representation plan to serve as a means to distribute anti-union propaganda. While Rockefeller had written that he was willing to let the workers decide whether management was their friend or their enemy, the firm's leadership was unwilling to trust workers to make up their own minds without offering additional input. In this way, the ERP became a means by which CF&I could make its case to employees in much the same way many anti-union employers do today under the National Labor Relations Act prior to a union vote. For example, consider Elmer Weitzel's attack on the UMWA at the January 19, 1922, joint conference of the Trinidad District, which management printed verbatim in the meeting minutes:

> I notice by yesterday's paper that our friends, [t]he United Mine Workers of America, have announced a policy, through President [John L.] Lewis, that regardless of what anyone else does, they are not going to take a reduction in wages. It is not for me to prophesy, but I believe Mr. Lewis is committing suicide. I can't see it any other way. Because the union has refused to accept a reduction, there are more than one hundred thousand of their members who have not worked since last July. The men in neighboring counties have taken their jobs.[70]

In a speech before the Walsenburg District meeting in September 1917, CF&I president Jesse Welborn warned against the danger "of placing in the hands of one or two, or even a handful of men the authority to call men out on a strike without a vote or expression by them of any kind." Welborn went on to declare, "We are determined to pursue the Industrial Representation Plan, develop the democratic features in it, and make it the guiding instrument, document, or agreement between our employees and ourselves, and we are just as determined not to recognize the United Mine Workers of America."[71] Welborn did not acknowledge any contradiction between the notion of democracy and management's refusal to recognize the UMWA, yet just two days before this speech he had received a letter from Starr Murphy telling him that the UMWA president had claimed that 90 percent of the company's miners were in the union.[72] The union members in the audience undoubtedly saw this contradiction more easily than Welborn did.

This policy may explain why the UMWA drastically changed its tactics a few years after the plan began. Instead of working within the ERP structure, the leadership of UMWA District 15 (which covered all of Colorado) eventually shunned the organization because it recognized that CF&I would never recognize any union under any circumstances. In 1920 the membership of District 15 banned members from serving as employee representatives, calling them a dual union like the Industrial Workers of the World. One representative was, in fact, later expelled.[73] What is ironic about the relationship between the Rockefeller Plan and the United Mine Workers of America was that the UMWA essentially gave up on organizing CF&I just as its workers began to recognize the limits of the plan. These frustrated CF&I workers had a much more sophisticated attitude toward the employee representation plan than the union did. As Selekman and Van Kleeck explained, one union-friendly employee representative "contended that union men should co-operate with and test the possibilities of the plan to its utmost limit and thereby seek to develop it into a union. The only alternative, in his opinion, was a strike for recognition of the union, and this he deemed hopeless with so powerful a company opposing [it]."[74] As was the case with the employee representation plans mandated by the National War Labor Board during World War I, union miners used their company union as a vehicle for the union cause. While this did not spur the conversion of non-union companies into union shops, as it would during the late 1930s (the necessary government support for such an effort did not survive the war), workers with ERPs invariably faced better terms and conditions of employment than workers who had no union at all.

THE 1919 MINERS' STRIKE

The year 1919 was full of strikes. One in every five American workers participated in a walkout that year, a record that, as David Brody has explained, "has never been surpassed, or even approached, in American history."[75] Among the most noteworthy strikes that year were the Boston Police Strike and the Seattle General Strike. It was also the year of the Great Steel Strike (discussed in Chapter 6) and a nationwide coal strike (discussed in this section). While the war was still in progress, the Wilson administration had forced employers in war and war-related businesses to at least talk to their employees regardless of whether they were organized into unions. Many unions, most notably the Amalgamated Meat Cutters and Butcher Workmen, used the war as leverage to organize. They knew meat packers were unlikely to fight hard when they were working at a frantic pace to supply the armed forces. When the war ended, the pressure from the Wilson administration and the need to fulfill war contracts ended as well. For this reason, many employers tried to take back wartime concessions they had granted to their unionized and non-unionized workers.

The causes of the nationwide coal strike were somewhat different. The UMWA and the companies that recognized it in its midwestern and eastern fields had signed an agreement in 1917 that dictated the terms and conditions of UMWA members' labor for the duration of the war. By mid-1919 the UMWA wanted to break the agreement and demand more concessions because of the employers' wartime profits and the miners' organizing success. However, the Wilson administration sided with the managers since, even though the fighting had stopped, there was still no peace treaty (and, in fact, there would not be a treaty for another two years). Although the president issued a public statement calling a potential strike "unjustified" and "unlawful," union miners in Indiana, Pennsylvania, and the rest of what was known as "the Central Competitive Field" came out in full force when the strike began on November 1, 1919.[76]

In Colorado, where no major mine operator yet recognized the UMWA, CF&I still went to great lengths throughout the war years to prevent its miners from joining the union. After all, John D. Rockefeller Jr.'s professed reason for having the plan was to prevent such events from happening. Nevertheless, tensions over UMWA recognition went straight to the issue of the plan's effectiveness and started long before the strike. As early as July 1917, President Welborn had published a letter in the *Industrial Bulletin* in which he offered to submit every grievance the union had, with the exception of union recognition, to arbitration by the Colorado Industrial Commission. "The Company is earnestly striving to protect and improve the interests of its employees," Welborn wrote. "Its loyalty to them cannot be questioned, and it naturally expects equal loyalty on the part of the employees toward the Company and the mutual interests of both Company and employees, as outlined in the Industrial Representation Plan."[77] After the war ended, however, employee loyalty disappeared. On the first day of the strike, 8,000 Colorado miners (about half the state's mining workforce) stayed off the job. Despite the existence of the ERP, CF&I employees responded to the strike in the same proportion as workers across the state. Every company mine except those around Walsenburg had to close down for lack of workers.[78] A majority of CF&I miners joined the strike even though the primary issues behind it had nothing to do with the union's national agenda. The only compelling plank of the UMWA's demands for CF&I employees was union recognition, which would have meant the end of the Rockefeller Plan.

Why did miners risk their jobs, especially with the benefits they had already gained under the auspices of the ERP and even though they might have turned to the ERP to improve their living and working conditions? Many miners thought the plan was part of the problem rather than the solution. An employee resolution that passed in many CF&I mines before the strike declared, "Said plan is not an agreement but a method of plain EXPLOITATION whereby the employees are compelled to work and exist under undesirable conditions. We have no proper protection, no voice or vote as to the conditions under which we shall work."[79] The union made this case in the field as well. "Gentlemen," said one UMWA organizer, "I hate these pets of the CF&I. You may have [a] good [superintendent] and your conditions may be good at Cameron, but they are not what they should be all over and we want to make them good all over and want a hundred percent organization."[80] In other words, the organizers at Cameron realized the advantages their miners had under the plan, but they made their case for an independent organization primarily to help others.

CF&I managers, for their part, worked from the assumption that the miners were happy under the plan, even after the strike provided a great deal of evidence to the contrary. They were no doubt led astray by bad intelligence on the ground, perhaps because Welborn automatically discounted the notion that anyone who joined the walkout actually opposed the ERP. As he wrote in the first issue of the *CF&I Industrial Bulletin* published after the strike, "Many of the workmen feared violence, claiming threats had been made against them by those promoting the strike, and in response to their petitions to the governor for protection state troops were placed in the different coal mining fields during the first days of November."[81] Ironically, the presence of troops at the mines only undercut Welborn's argument, as the troops could presumably have ensured any strikebreaker's safety. Nevertheless, enough miners stayed out to keep CF&I's production below 50 percent of its normal capacity until the very end of the strike.[82] This is yet another indication that management could not trust workers to express themselves truthfully about labor issues.

In Colorado, the strike lasted only eleven days. Colorado District 15 president George Johnson planned a second strike on November 20, 1919, in response to an allegation that seventy union members had been denied employment, but a court order stopped the strike from occurring. According to management, thirteen of these men were denied employment for using unlawful means to deny others employment during the strike, but it produced no evidence or even details when Ben Selekman and Mary Van Kleeck asked for them a few years later. Many of the others not rehired were union leaders who all concluded that they were denied reemployment solely for that reason. Selekman and Van Kleeck described the case of Mike Glad, the financial secretary of the union at Tollerville: "[Glad] was a very active union man. He had been criticized for this by the superintendent, who, because of it, had refused to assign a good home to him. Tollerville had also closed down on September 22 because of the steel strike. When it re-opened some time in December, Glad returned for his work. He was refused re-employment."[83] When managers testified about their refusal to rehire Glad before the Colorado Industrial Commission, they said he made threats when soliciting members, but they were only relying on hearsay.

The Colorado Industrial Commission investigated the charge of discrimination against union members and exonerated CF&I of the allegation. Nevertheless, Fuel Department manager Elmer Weitzel admitted to a U.S. Senate committee in 1920 that seven miners who had made "inflammatory talks . . . were taken by the committees [of employee representatives] to the outside of the camp and told not to come back."[84] Perhaps this explains why the Industrial Commission believed that the employee representatives had acted independently of management. After the walkout, management forced returning miners to sign this statement: "As an employee of the Colorado Fuel and Iron Company, I know that it is operated as an open shop under the Plan of Representation of Employes of which I have received a copy. I will cooperate in maintaining the rules and agreements relating to my service and the laws of my State and Country."[85] The company suspended this effort, however, when the UMWA complained to the Colorado Industrial Commission.[86]

"I can quite understand why you were led to try the card plan," Rockefeller wrote to Welborn after the change of policy:

> Of course, it never occurred to you in adopting the plan that the company might thus lay itself open to being put in a position where the strikers would find it possible to have their position sustained by the public authorities, even unofficially. Having made the experiment, you now know the advantages and disadvantages of the system and will be better able to determine what is wise in regard to the continuation or abandonment of the policy from now [on].[87]

Rockefeller did not chastise Welborn in this instance for violating a cardinal principle of employee representation. Faced with evidence that the concept did not work the way he and King had imagined it, he merely noted that violating the concept in this instance was unwise.

Rather than fix the plan to prevent further unrest, CF&I management made excuses. As one executive wrote (and John D. Rockefeller Jr. seconded), "It is true that the Industrial Representation Plan did not prevent the men from going on strike. The conditions . . . were such that it was practically impossible that this plan, or any other system, could have prevented this walkout. The manner in which the strike was called and the admission of the leaders that there were no grievances, should be sufficient evidence that the Industrial Representation Plan removed all just causes for grievances."[88] The desire for recognition of an independent union was not a "just" grievance in the minds of the men who ran the Colorado Fuel and Iron Company. They were incapable of putting themselves in their miners' shoes and seeing that the miners wanted a kind of independence the ERP could not provide.

For this reason, a plan originally conceived as a way to end all labor conflict did not work as intended. The smaller, shorter strikes among

THE ROCKEFELLER PLAN IN ACTION: THE MINES

CF&I coal miners in 1921 and 1922 underscored support for the UMWA among the company's employees as well as widespread distaste for the Rockefeller Plan among rank-and-file miners. By the late 1920s, management knew that if it gave miners enough freedom to decide for themselves whether labor or management was their best friend (Rockefeller's phrase), most miners would choose labor. This attitude, obvious to all observers after the statewide 1927–1928 strike led by the Industrial Workers of the World, would set the stage for management's decision to abandon the plan when the UMWA began reorganizing the Colorado coal fields during the early years of the New Deal.

Chapter Six

THE ROCKEFELLER PLAN IN ACTION: THE MILL

> It is easy to have harmony if we give the other fellow everything he asks for.
>
> —J. F. CHAPMAN, SUPERINTENDENT, MINNEQUA WORKS, 1924[1]

On September 25, 1919, President Woodrow Wilson delivered the final speech of his nationwide tour in support of the Treaty of Versailles and the Charter of the League of Nations contained therein at Pueblo's new Civic Auditorium. Among those in the audience were striking steelworkers from the Colorado Fuel and Iron Company (CF&I) who had left their jobs three days before. Among the least-remembered parts of the Treaty of Versailles is a series of clauses relating to international labor rights. Most important among them was the one that created the International Labor Organization, which still exists today. While these clauses received little attention at other stops on his tour, Wilson did not ignore the issue in Pueblo:

> Reject this treaty and this is the consequence to laboring men of the world, there is no international tribunal which can bring the moral judgments of the world to bear upon the great labor questions of the day.
>
> What we need to do with the labor questions of the day is to lift them into the light—is lift Americans out of the haze into the calm spaces where men look to things without tragedy.[2]

This is exactly what John D. Rockefeller Jr. had intended the Rockefeller Plan to do—prevent tragedy by facilitating communication over labor issues at CF&I. Rockefeller equated his ideals with President Wilson's in a speech before the U.S. Chamber of Commerce in Atlantic City, New Jersey, less than three months later.[3]

Like many working-class Americans, steelworkers at CF&I responded very favorably to Wilson's call for a League of Nations. In a series of resolutions striking steelworkers presented to Wilson shortly after he arrived in Pueblo, the strike committee from the mill linked the struggle for democracy in the workplace specifically to Wilson's ideals:

> Be It Resolved. That the six thousand five hundred peaceful and law abiding workmen in Pueblo fighting a forced battle against industrial autocracy, do hereby take occasion to assure you of their abiding faith in the principles of democracy as enunciated by you, and to protest most vigorously against the forcible continuation of paternal and undemocratic institutions in vogue to befool the employe and deprive him of his right to collective bargaining.[4]

The "paternal and undemocratic" institution the strike committee had in mind was, of course, the Rockefeller Plan.

The strikers also used Rockefeller's rhetoric against the company during the walkout. "If John D. Rockefeller puts into effect here in Pueblo the suggestions he made at the round table at the Labor Conference in Washington [which was going on at about the same time], the strike would be settled at once," explained John Gross, the chair of the strike committee, to *The Denver Post*. "[B]ut for some reason he ignores us."[5] The steelworkers, in effect, had called Rockefeller's bluff.

Despite the presence of the Rockefeller Plan—in fact, specifically because of it—many CF&I workers joined the nationwide steel strike organized by the National Committee for Organizing Iron and Steel Workers. Following a failed conference between CF&I president Jesse Welborn and the strike committee, the chair of the committee accused Welborn of attempting "to jam the plan down the men's throats."[6] (The committee had asked to talk to John D. Rockefeller Jr. He declined.) As with the miners' strike that same year, the strike at the Minnequa Works in 1919 suggested that the plan had failed in its core mission of pacifying labor. While the possibility of unflattering comparisons to the United Mine Workers of America (UMWA) could rightfully be blamed for the miners' decision to reject the Rockefeller Plan, CF&I's steelworkers should have been grateful for their "company union," since the steel industry had no viable independent union to which it could have been compared. Nevertheless, many preferred the independent union that did not exist to the employee representation plan (ERP) that did.

The strikers' hostile response to the plan should not have surprised management. Significant opposition to the plan existed among steelworkers from its inception at the Minnequa Works. When management first sent the plan to workers for approval in 1916, 863 steelworkers, or 27.1 percent of those who bothered to vote, voted against it. At the very first meeting, steelworkers complained about CF&I's failure to raise wages to match the high cost of living during the war years, as well as their inability to impact wages under the terms of the plan. This proved to be a deciding issue in recruiting steelworkers to the strike effort.[7] By the time the strike started, four separate union locals were already active in the steel mill.[8]

President Wilson failed in his efforts to get the U.S. Senate to ratify the Treaty of Versailles (largely because of a debilitating stroke he suffered on the train a few hours outside Pueblo).[9] Similarly, the steelworkers failed to get CF&I to practice the kind of democracy they envisioned rather than the kind imagined by John D. Rockefeller Jr. Pueblo steelworkers did not strike as often as their colleagues in the mines. Besides having no independent union they could rely on for help, the steelworkers had no chance of finding another job in steelmaking anywhere near Pueblo. Nevertheless, outside of a few of the most highly skilled workers in the mill, the response to the plan within the Minnequa Works was largely negative. Even the skilled steelworkers expressed dissatisfaction with the plan through the ERP itself. Mark M. Jones, who reviewed the industrial relations situation at the steel mill in 1925, described it this way: "[The workers] say that it is a case of 'every man for himself' and they should 'get theirs.'"[10] This was not what John D. Rockefeller Jr. and Mackenzie King had in mind when they designed the plan, but the development of such an attitude was perhaps inevitable in an industry where skilled workers like Andy Diamond still dominated the production process and held tightly to their comparatively high-paying jobs for long tenures. This sense of job security encouraged such workers to speak their minds in ways workers lower on the job ladder could only imagine.

THE BENEFITS OF HAVING A VOICE

Like the representatives in the mines, workers who participated in the Rockefeller Plan at the steel mill often pressed their grievances in extremely strong terms. In 1925, for example, an employee representative from the Minnequa Works stood up at a meeting, accused management of being incompetent, and threatened to inform stockholders of his opinion.[11] Such outbursts cannot be reconciled with traditional stereotypes of pliant workers manipulated by management through sham organizations. By using the Rockefeller Plan to express their individual voices, steelworkers forced upper management to make the plan work for them. As the secretary to former president's industrial representative B. J. Matteson explained in 1928, some representatives literally forced the ERP to work "in spite of [a] lack of sincerity on the part of at times dominating influences in the local management towards the Plan, which was quickly sensed by groups of workers."[12] By that time steelworkers had made considerable progress through using the plan for their own ends. Gaining the eight-hour day six years before most of the rest of the steel industry and earning higher wages than those paid to most steelworkers, even in the East, were two examples of that progress.

Victories such as these sparked a backlash among middle management, primarily superintendents and foremen. Many of the complaints the steelworkers made were lodged directly against foremen. For example, in July 1924 foreman George Plute struck steelworker Merle Friar after Friar disobeyed a direct order from Plute to inspect some rails. Management suspended Plute for a week but did not discipline Friar. Rather than leave well enough alone, Friar appealed Plute's suspension through his employee representative, suggesting that Plute should have been fired. That did not happen, but the situation illustrates both the steelworkers' belief in their immunity from discipline and their willingness to go over the heads of their immediate superiors.[13] Indeed, employee representatives in the steelworks often threatened to go over management's collective head during negotiations by appealing to President Jesse Welborn or even to John D. Rockefeller Jr. Occasionally, they actually did so.[14] "I was told by the representatives about the 5th of June 1928," Minnequa Works superintendent Louis F. Quigg remembered ten years later, "that I would either live up to that plan or I wouldn't work in this plant."[15] Steelworkers made their voices heard in this manner because they built a record of success in getting what they wanted from management.

They achieved these victories by developing a broader range of demands than management wanted to offer them. "It is significant," concluded Ben Selekman, "that once having been granted representation, the employes of the Minnequa Works desired a voice in all decisions affecting conditions of work."[16] The company's decision to shorten the working day to eight hours (discussed later) was probably the steelworkers' greatest victory under the plan, but representatives won other victories as well. For example, at the June 26, 1918, meeting, steelworker John Burkhardt "presented a preliminary constitution of a sick and accident cooperative insurance plan, which had been largely taken from the Carnegie Plan at [the] Homestead [Steel Works]." When that plan was tabled for further investigation, "Mr. Burkhardt said that he would like to hear some suggestions from the superintendents or foremen for the benefit of the plan. [Gotlieb] Schultz [another employee representative] mentioned that the company representatives were spectators or wallflowers; that the department didn't seem to try to help or assist in this work. That they discussed things among themselves, but didn't seem to speak up at meetings."[17] When representatives continued to raise the insurance issue at subsequent meetings, management instituted a mutual benefit association in 1920, slightly modified from the original proposal to comply with Colorado insurance law. This employee benefit would never have existed if worker representatives had not pressed the issue.

The same is true of the YMCA building. At the first meeting of employee representatives at the steel mill in 1916, Representative F. Joseph Loeffler "expressed his conviction that there was a great need for a club at the Steel Works and the hope that Mr. Rockefeller might be interested in helping to provide a building for such a club."[18] Later, the company announced plans to build the Steel Works "Y."[19] While the company publicized the Rockefeller-funded building as a gesture of goodwill by the Rockefeller family, it was actually one in a long line of concessions management made to CF&I employees.

THE EIGHT-HOUR DAY

Of all the concessions CF&I offered its employees, the decision to put its steelworkers on an eight-hour workday received by far the best reception. Despite more than occasional differences of opinion, CF&I steelworkers spoke in one voice in support of shorter working hours. "If the company ever attempted to reestablish the twelve-hour day," one employee representative told Ben Selekman, "the men would just spontaneously walk out and strike and I would be one of their leaders."[20] CF&I management first gave its steelworkers the eight-hour day in November 1918. (Most of the company's coal miners received the eight-hour day sometime around February 1913.)[21] Although CF&I was not the first firm in the steel industry to do this, it was still far ahead of most other steelmakers (many of which did not change their policies until pressured by the Harding administration in 1923). The inspiration for the eight-hour day at CF&I was U.S. Steel's decision to introduce the basic eight-hour day for its employees in an effort to keep its mills running smoothly because of the war.[22] The basic eight-hour day meant employees were paid overtime for all work beyond eight hours on any day. The actual eight-hour day meant they worked eight hours. By granting its workers the actual eight-hour day, CF&I made a more significant concession and helped drive the entire steel industry toward doing the same.

World War I formed the backdrop of the debate over hours in Pueblo. Orders had increased greatly because of American mobilization efforts, and management wanted the workers to be happy and productive. That is why U.S. Steel agreed to pay overtime. America's steelworkers, for their part, were not willing to work themselves into a state of exhaustion. Therefore, they willingly sacrificed some pay for more free time. U.S. Steel's workers accepted a 10 percent wage increase and eight hours of work each day instead of a $16^{2}/_{3}$ percent increase for twelve hours of work each day.[23] Yet when CF&I management called a meeting to discuss how to introduce the basic eight-hour day at the Minnequa Works, employees requested the actual eight-hour day at only a "slightly higher" wage.[24] Representative A. H. Lee "argued that the men did not put forth their best efforts when they had to stay at their jobs every day for so long a period, and was of the opinion that they would produce as much, if not more, when working eight as when working twelve hours."[25] Afraid of alienating management, no other employee representatives supported Lee at the meeting. In response to Lee's suggestion, President Welborn (who attended the meeting) expressed support for the actual eight-hour day in theory but argued that his hands were tied unless U.S. Steel and other competitors instituted this costlier concession first.[26]

Despite their failure to speak up when the proposal first arose, employee representatives persisted in their fight for the actual eight-hour day. The seventh division at the steel mill (which represented the carpenters, pattern makers, machinists, and other skilled laborers) continued to push for the actual eight-hour day after management first rejected the idea. Many of these workers were members of craft unions (even if CF&I management refused to recognize those unions in Pueblo) that represented specific trades in which the actual eight-hour day had become increasingly common. Management agreed to the eight-hour day for these workers first. When other steelworkers heard about this arrangement, they "declared themselves emphatically" in support of it for themselves as well. Management quickly gave in across the board.[27] The actual eight-hour day began at the Minnequa Works on November 3, 1918, even though U.S. Steel had not yet shortened its hours. Employees willingly took a cut in their total pay (the 10 percent wage increase did not make up for the wages workers lost in the missing four hours each day) in exchange for shorter hours. The actual eight-hour day was a great victory for steelworkers, especially as the Memorandum of Agreement that accompanied the ERP in both the mines and the mill stated that determination of the number of hours worked was the sole prerogative of management. Whether as a result of anger at sacrificing any wage revenue or out of hope that they would be able to extract future concessions, it was probably no coincidence that CF&I steelworkers formed the Lafayette Lodge of the Amalgamated Association of Iron, Steel and Tin Plate Workers the same day the actual eight-hour day went into effect.[28]

Even if the eight-hour day inspired some steelworkers to press for more concessions, it did not take management long to recognize the benefits of this new policy. In a letter to the Federal Council of Churches explaining the results of this change, President Welborn wrote:

> The trend of production per man-hour, with unimportant exceptions, has been upward since the adoption of the 8-hour day; and in every department of our steel manufacturing operations, from blast furnace to the wire mill, our production per man-hour is now greater than it was when all of these activities were operating on the 12-hour shift. Comparing these results of the last few months with periods of similar

production when basic rates were 10 percent lower than current rates and the working time 12 hours per day, we find that almost without exception our labor cost per ton is lower than in the earlier periods.[29]

Management would not have gained these improvements in production if it had not allowed steelworkers an avenue to express themselves and then actually listened to those demands. Besides the increase in production, management also benefited from the cut in total worker compensation (compared to what it would have paid out had it paid overtime) that accompanied the change.

Despite the economic benefits, management became uncomfortable about continuing the eight-hour day when most of its competitors resisted the change. When U.S. Steel and other eastern mills finally adopted the actual eight-hour day in 1923, wage rates changed for those companies as labor and management shared the cost of shorter hours. After this change, CF&I management sent a committee of employee representatives to the East to survey the situation. After they returned, the workers demanded a twelve-cent-an-hour increase and got it, but management wanted employees to work the same schedule as those in the eastern mills, meaning some employees would have to work twelve hours a day.[30] The employees again rejected the suggestion. During this 1923 dispute, the employee representatives passed a unanimous resolution that read, in part:

> We maintain that the 12-hour day in the steel industry works an unreasonable hardship on the employees; that it is physically injurious to the employees; that it interferes with family associations and, for these and other reasons, is opposed to the public interest.
>
> We maintain that regardless of this committee's convictions to the contrary [referring to the management members of the Joint Committee on Industrial Cooperation, Conciliation, and Wages, which made the proposal to re-lengthen the workday], the 12-hour day is of itself an injury to workmen, physically, mentally and morally.
> . . .
> We flatly deny that the workmen prefer the longer hour[s] because it permits larger compensation per day.[31]

No response to this resolution from management is recorded in the record, but the fact that it published this response in a supplement to the *Industrial Bulletin* says a great deal about management's willingness to tolerate dissent. The dissent was successful, since nobody at CF&I went back to working longer hours.

WAGES

Steelworkers gained a great deal through the employee representation plan, but their most difficult battle was the one for increased wages. The coal industry had a price structure based on local markets, but the steel industry had a national price structure based on a concept called "Pittsburgh Plus." Pittsburgh Plus was instituted by the U.S. Steel Corporation early in its history to guarantee that steel generated in Pittsburgh would be competitive everywhere in the United States. In other words, CF&I had no cost advantage in its primary sales area, even though it was virtually the only steel mill in the West until World War II. This meant that CF&I's steel division was very price sensitive and did not want to increase wages unless its competitors, especially U.S. Steel, did so first.

Despite a clause in the Memorandum of Agreement similar to that in the mines (see Appendix 1), requiring that "similarity of rates" with "companies whose products are sold in active competition with products of the Colorado Fuel and Iron Company . . . shall be maintained," from the very first plan-related meeting in the steel mill, steelworkers continually pressed the company to raise their wages. In his first speech to the new employee representatives, President Jesse Welborn appealed to his workers' self-interest in an effort to hold off complaints about wages. "I feel and know that our interests are mutual," he told the representatives, "that it will do the Company no permanent good to force the workmen to a lower wage than they ought to take, and that the Company is obliged to pay a higher scale of wages than its trade will justify, [yet] no permanent good will result to workmen." (This plea alone demonstrates that management did not hold all the power in the plant.) Yet minutes after Welborn finished his speech, employee representative Frank Van Dyke, "in a short address, touched upon the reduction made in 1914 in the wages of employees in the Open Hearth Department, which he said, on account of business conditions at the time was justifiable, but in the view of the activity of business at present, he felt that the scale of wages in effect was unjust."[32] This was just the first of many requests for wage increases made under the auspices of the plan. In 1921, the Joint

Committee on Cooperation and Conciliation became the Joint Committee on Cooperation, Conciliation, and Wages at the workers' request.[33] This meant that management succumbed to employees' pressure and actually let them bargain over wages, contrary to what the plan originally stated. Management's efforts to use the plan to take wages off the table had failed in the face of resistance from employee representatives.

CF&I steelworkers also faced the issue of how often they worked when considering their wages. As explained in Chapter 4, steel companies tended to take as many orders as possible and run their mills at full capacity for a limited time, then shut down for extended periods. In most plants, this practice stopped during the war when the industry essentially "discovered" the negative impact of employee turnover.[34] At CF&I, the long shutdowns continued into the 1920s. During the economic downturn in 1921, operations at the Minnequa plant slowed to almost nothing. As a result, the company laid off workers and cut back the hours of those who remained. At that point, management demanded a reduction of wages in exchange for calling more men back to work. "Conditions were very bad," an employee representative wrote Ben Selekman in a letter published in his 1924 study of the plant. "Most of the men had not worked for several months and we, the representatives, knew the plant would not start until the proposed reduction in wages took effect."[35] After spending two day consulting with their fellow workmen, the employee representatives countered with a 15 percent cut. Management accepted the offer on January 16, 1921.

In 1921, steelworkers had greater success in increasing their wages. Indeed, compared with other workers in the rest of this non-union industry, they won more concessions than anyone, including other steelworkers laboring under employee representation plans.[36] This record of success required patience and persistence. An employee representative who had held office for five years told Ben Selekman: "I would state that with very few exceptions, wage cases taken up with the local management previous to [1921] met with scant consideration."[37] Yet because management's policy was "to avoid differences with the workmen at any expense," by 1925, wages at CF&I were on average 31 percent higher than those of comparable workers in mills back East.[38] That did not stop the steelworkers from demanding more, however. That same year, 1925, management attempted to cut by 20 percent the rate paid per ton to employees working in the rail mill. When the steelworkers refused to accept the cut, management brought the matter before the Joint Committee on Industrial Cooperation, Conciliation, and Wages. To persuade them of the need for the cut, it again sent a delegation of employee representatives on a tour of eastern facilities. After they returned, the workers proposed a cut of only 5 percent, and management accepted.[39] All this happened under the auspices of a plan that was not supposed to cover wages as originally written and despite the steel industry's nearly universal attitude of maintaining low labor costs at any price into the 1920s.[40] Because workers at CF&I could use the Rockefeller Plan to lobby for higher wages, the company became a notable exception to this rule.

SAFETY IN THE MILL

Safety should have been one issue on which steelworkers and their employer generally agreed, since both labor and management stood to benefit from its successful implementation. Had effective safety measures been in place, workers would have suffered fewer debilitating (and sometimes fatal) injuries on the job. At the same time, management would have saved money because the fewer the number of accidents at the mill, the less it would have contributed to Colorado's state workmen's compensation fund. Fewer injuries would also have meant increased worker productivity, as fewer employees would have missed work because of injuries. Unfortunately for both labor and management, it did not work out this way.

When John D. Rockefeller Jr. visited the plant in 1915, he "heard the shout of 'low bridge' and just escaped in time to avoid getting a wallop." According to the *Chicago Tribune* reporter who accompanied him on the tour, "A dozen times this afternoon a misstep might have meant his death."[41] Despite the safety operations instituted under the plan, both the number of accidents and the amount of time workers lost in the steelworks as a result of accidents increased between 1919 and 1923. In fact, the Minnequa Works had twice as many accidents as other plants in the industry.[42] These figures are particularly noteworthy because the eight-hour day should have decreased accident rates, since workers would have been less tired than when they were working twelve-hour days. Accident rates at other mills diminished when the working day was shortened, but safety problems at the Minnequa Works persisted. In 1926 the Joint Committee on Safety and Accidents lamented that its "annual report is in some measure a record of disappointment, because the results achieved appear so pitifully small when compared to the efforts put forth."[43] In 1927, even management acknowledged that the company "seemed to be progressing backwards" on the issue of safety.[44]

As was the case in the coal mines, safety at the Minnequa Works was a point of tension between labor and management, mainly because each side blamed the other for being the primary cause of accidents at the mill. In 1923, employee representative Warren Densmore wrote in the *Industrial Bulletin*:

> The three reasons most men are injured are, in my opinion, and in the order of their importance:
>
> 1. Management at fault.
> 2. The thoughtlessness of the workmen.
> 3. The carelessness of the workmen.
>
> Management will get results in safety work in direct proportion to what is put into it.[45]

Not surprisingly, Harrington Shafer, the safety director at the Minnequa Works, had a different opinion of where the fault for most accidents lay. "According to a common classification of accidents," he wrote in the same issue of the *Bulletin*, "the far greater number are caused through carelessness."[46] More specifically, Shafer had claimed earlier that year, "You can go through the records of years past and you will find that eighty or more percent of all accidents could not have been prevented mechanically."[47] Even workers who praised the company's safety efforts to Ben Selekman complained that management did not implement safety instructions fast enough and that the safety procedures at the non-union U.S. Steel Corporation were more effective. In response to such criticisms, management claimed it could not afford the expense that modern safety equipment and machinery would require.[48]

The best indicators of the tension over safety did not surface in the minutes of meetings held under the auspices of the Rockefeller Plan but rather in meetings held in Denver with government officials. During the

THE ROCKEFELLER PLAN IN ACTION: THE MILL

life of the ERP, countless CF&I workers faced off against management in workmen's compensation hearings before the Colorado Industrial Commission, squabbling over blame for their accidents, the size of their disability pensions, or both. No case illustrates this tension at the steel mill better than that of Earl Ostrander. On August 18, 1926, Ostrander was "blown out of a slag pit" at the Open Hearth. The accident burned out his eyes, resulting in total blindness. It also burned his legs, chest, arms, and sides and left scars on his face. Ostrander took his case to the Colorado Industrial Commission and received a twenty-five-dollar-a-month pension. As CF&I president Jesse Welborn admitted, "No fair minded person can say that the amount fixed as compensation for such an injury [as] you have suffered will, even in a small measure, reimburse you for the loss of eyesight, nor would any sum, no matter how large, compensate one for such loss."[49] Between 1929 and 1935, Ostrander, in letters dictated to his mother, made appeals for more money not only to John D. Rockefeller Jr. but also to Rockefeller's minister (Harry Emerson Fosdick of Riverside Church in New York City), to Nelson Aldrich (Rockefeller's brother-in-law) of what was then the Chase National Bank, and even to Rockefeller's wife, Abby.

To say that the case became an embarrassment for everyone connected to the company except Ostrander would be a massive understatement. The picture shown here of Ostrander wearing his sandwich board was no doubt intended for the local newspaper. The most interesting part of his message for purposes of this study is the way Ostrander linked his fate, both on that board and in his letters, to the Rockefeller Plan. "This Plan [Rockefeller Plan] is supposed to protect the employes, their rights, their jobs and insure a fair and square deal, for each and all," he wrote in his appeal to Aldrich. However, "All the Officials did was to offer me the same scanty compensation and the useless blind education again without any job. The consideration given me by the C.F.&I. Company Officials is a disgrace to this Company and the Rockefeller Plan."[50] Clearly Ostrander, a skilled native-born American and a twelve-year veteran in the plant at the time of his accident, had gained a sense of entitlement tied to the ERP. When his accident forced him to realize that management held all the cards in the employment equation, his frustration became palpable.[51]

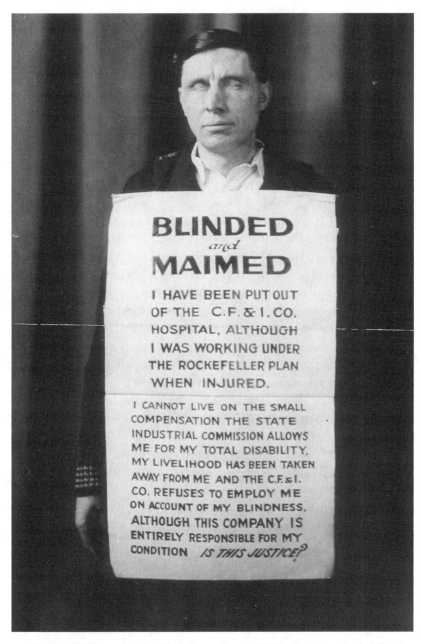

6.1. *Former CF&I steelworker Earl Ostrander, ca. 1929. Courtesy, Rockefeller Archive Center, Sleepy Hollow, NY.*

THE 1919 STEEL STRIKE IN PUEBLO

The 1919 steel strike surprised observers throughout the industry. The infamous 1892 Homestead Lockout had broken the back of the once strong Amalgamated Association of Iron, Steel, and Tin Plate Workers; therefore, the steel industry had been almost entirely free of strikes since the new U.S. Steel Corporation beat that union again in 1901. The National Association of Iron and Steel Workers, led by Chicago American Federation of Labor leader John Fitzpatrick and future Communist Party head William Z. Foster, managed to infuse life into the industry's beleaguered workers by concentrating its efforts on the most exploited employees—immigrants who generally held the least-skilled jobs in the mills. The workers went out on strike in September 1919. The Amalgamated Association's decision to cross the picket line in early November, less than two months after the strike began, crippled the efforts of the union nationwide. With the mills back at full production, the National Committee for Organizing Iron and Steel Workers called off the strike in January 1920.[52] In Pueblo, however, the strike had its own rhythm because of the presence of the Rockefeller Plan. The plan both helped cause the strike and served as a tool to help end it.

Less than a week before the national walkout began, the employee representatives at the Minnequa Works met with management for their regular quarterly conference. Even though the National Committee for Organizing Iron and Steel Workers had already called a nationwide steel strike for five days later, neither side mentioned unions or strikes at the meeting. In fact, as CF&I resident Jesse Welborn described it in a private letter to John D. Rockefeller Jr. shortly after the strike began, "[T]here was no indication that our employes contemplated a strike, or that they desired us to make any radical or material change in our policy."[53] Welborn received no indication of a strike because national strike leaders and the leaders of the local organizing campaign were not planning to strike when they met with management. However, at a meeting the next day, CF&I steelworker Johnnie Jones put forward a motion calling for the union to go out in sympathy when their colleagues around the country struck. To everyone's surprise, the motion passed overwhelmingly.[54] A committee of five employees called on steelworks manager F. E. Parks and presented their demands, which centered on the abolition of the

THE ROCKEFELLER PLAN IN ACTION: THE MILL

Rockefeller Plan and recognition of the union.[55] They did not bother to notify the Colorado Industrial Commission as required under the provision of the 1915 law that imposed thirty days' notice before any strike or lockout, even though not doing so left the union open to the possibility of heavy fines.[56]

Most of the demands the national committee made to employers back East did not apply to CF&I. Under the auspices of the ERP, management had already made many of the concessions for which steelworkers at other firms were striking even before the strike began. "We had wages. We had hours," one former striker later remembered.[57] According to Welborn:

> The discussion at that meeting [five days prior to the strike], and with the committee of five previously referred to at another meeting this morning, brought out free acknowledgement by a considerable number of the representatives and by the committee that their working and living conditions under the operation of the Industrial Representation Plan had greatly improved and were now much better than prevailed in Eastern plants[,] that the treatment received at the hands of the Company was fair and that there had been no discrimination against men on the part of the Company officials because of union activities.[58]

Other workers told management that organizers for the national committee had informed them that instead of striking, steelworkers in Pueblo would be called on to give "moral and perhaps their financial support."[59] The representatives who presented the union's demands to Welborn predicted that most steelworkers would ignore the strike call.[60] Indeed, not even the union leaders themselves knew whether the strike in Pueblo would be as popular as it was in the more autocratic mills in the East.

The union claimed 80 percent representation in the mill before the strike.[61] Approximately 95 percent of Minnequa Works employees failed to report to the plant on September 22, 1919, the first day of the walkout.[62] As *The Pueblo Chieftain* reported regarding a statement put out by the strike committee that day, union leaders thought the response to the strike call "was a complete success and an emphatic denial that [CF&I steelworkers] were satisfied with the Rockefeller Industrial Plan."[63] Unlike some mills in the East, there was virtually no violence at any CF&I property during this nationwide strike.[64] Management attributed the lack of violence to the fact that labor's only significant grievance was the company's failure to recognize the union, but it recognized that this might change if CF&I attempted to operate the mill with scab labor.[65] Therefore, CF&I waited out the dispute.

While the idea of management recognizing an independent union at this juncture might seem bizarre in retrospect, some of the employee representatives at the mill were already union members, and management knew it. Furthermore, CF&I had signed a contract just a few months before with the independent Brotherhood of Railroad Trainmen and the Brotherhood of Locomotive Firemen and Enginemen for its Colorado and Wyoming Railway line. Railway workers dealt directly with steelworkers every day in the yard outside the mill. In fact, employee representatives from the steelworks served as intermediaries between the company and the brotherhood during the negotiations, since they feared a railway strike would lead to management shutting down the mill. The decision to recognize the railway workers' organization demonstrated that CF&I management was willing to bargain with outside unions under certain circumstances. The strike committee made this an issue in the strike, addressing Welborn in an open letter: "Now Mr. Welborn you signed an agreement with your employes on the Colorado and Wyoming railroad last December. These men number something over two hundred. Why can't you do business with six thousand of your men at this time?"[66] The answer was, of course, that such a policy would have gone beyond the acceptable limits of democracy at CF&I and the committee knew it, which is why it made the Rockefeller Plan the central issue of the strike.

CF&I management mistakenly thought the presence of the ERP in its mill would protect the company from the wave of organization that swept across the steel industry that year.[67] This misplaced sense of security was obvious before the strike began. "All the men on the outside on Monday will be yours," explained Welborn to an employee representative (who also happened to be a union leader) prior to the strike, "and all those on the inside will be mine."[68] By Welborn's own standard, therefore, the strike reflected very poorly on management's labor policy.

Nevertheless, the Rockefeller Plan still helped to end the strike it had helped instigate. About ten days after the strike began, employee representatives formed a back-to-work organization, consisting of 45 members, to end the strike.[69] On October 9 a statement in the local newspaper revealed that a back-to-work organization had been formed. A story three days later reported that the organization had 2,655 members.[70] Officially, the strike lasted until the mill restarted operations on December 17. Fear of serious violence explains why management used a back-to-work movement to slowly build momentum for restarting the plant rather than doing so as soon as possible. The plant had approximately a quarter of its normal workforce when it restarted. The national strike formally ended on January 8, 1920. According to William Z. Foster, local union leaders had to beg their men to go back to work; in fact, only 1,500 of 6,500 men had done so. The reason for such resistance, claimed Foster, was the Rockefeller Plan. "This worthless tyrannical arrangement the men could not tolerate," he wrote in his memoir of the strike.[71] Despite such distaste for ERPs, the strike did not help the cause of unionization in the steel industry. According to the Colorado Industrial Commission, the strike cost steelworkers $2 million in lost wages.[72] Nationally, Amalgamated Association membership dropped from 31,500 in 1920 to only 4,700 at the beginning of the New Deal. The union lost every labor dispute it faced during those years and did not manage to organize a single new facility.[73]

THE LIMITS OF INDUSTRIAL DEMOCRACY

In a letter printed in the *Industrial Bulletin* following the strike, CF&I president Jesse Welborn expressed the same surprise about the steel strike that he had to John D. Rockefeller Jr. shortly after the strike began. "No dissatisfaction with working conditions or wages had previously been expressed by employees," he wrote.[74] While Welborn may have believed the firm's steelworkers were content, the fact that so many struck clearly demonstrated that this was not the case. This reality, in turn, demonstrates that most steelworkers were unwilling to express their unhappiness through the vehicles set up under the Rockefeller Plan, an extraordinary failing for a plan designed to get both labor and management to recognize their mutual interests.

With no viable independent union to turn to after the strike, CF&I steelworkers had to use the ERP the way they would have used an independent organization had there been one in their mill. For example, between 1921 and 1925, employee representatives made multiple requests for longtime CF&I workers to be given paid vacations. Management consistently

THE ROCKEFELLER PLAN IN ACTION: THE MILL

put them off. On December 30, 1921, for example, President Welborn did not explicitly deny one such request but stated that the company would not implement a vacation policy at that time.[75] This, apparently, was a common management strategy. Earlier, Representative W. D. Turner had complained that "many times questions have been taken up by various committees . . . and these questions have been referred for an answer or consideration at some future date. Many times we never hear about what action is taken on the recommendation unless we bring the matter up."[76] This was a strong reminder from management to employee representatives that it held most of the power in their bargaining sessions.

Management may have put off the steelworks employee representatives because it realized that the response to many of their grievances would only upset them. Consider a complaint about the quality of the water the men in the mill drank, leveled during the joint conference on June 6, 1919. The aggrieved representative was Warren Densmore: "[T]he artesian water we are drinking does not agree with the men. If you will take a bucket of that water and let it stand there is a skum [sic] that forms on the bucket—it is either iron or something else, but it will eat the bucket, and if it is strong enough to eat through a bucket, you can imagine what it will do to a person's system." The company doctor, who was at the meeting, claimed the water had been tested and had "just a grain of iron per gallon." He blamed whatever symptoms the men had after drinking the water on "ice water colic," a condition supposedly derived from drinking too much water in extreme temperatures.[77] The inside of a steel mill is always extremely hot, but there is no need to explore the merits of this diagnosis to understand how the employees must have felt. Who would drink buckets of scummy water under any circumstances, let alone at a time when proper hydration mattered most? The plant manager's reply to the complaint must have only made the situation worse. "We had a refrigerator company make bids on installing a water system throughout the steel plant," he told the assembled representatives, adding that "the bids were entirely out of reason and we would not consider them."[78] Management had told workers in the steel mill as early as December 1917 that it was working on the water quality problem, but two years later the company had still done nothing.[79] The issue was not reported on again in available ERP minutes. "One of the most active employee representatives" described the cumulative effect of such slights to Ben Selekman: "It

THE ROCKEFELLER PLAN IN ACTION: THE MILL

is becoming more and more evident that practically every dispute brought before officials for settlement assumes the nature of a contest. Very little is decided without a lengthy argument. If this antagonistic attitude persists, the employees will be compelled to seek other means to secure justice."[80] By "other means," he no doubt meant an independent trade union.

Of all the controversies that arose under the plan, seniority is one of the best issues to demonstrate the similarity between the way the Rockefeller Plan operated in the mill and how an independent trade union might have handled similar situations. As Michael John Nuwer has explained in reference to the steel industry in general, "After the union's expulsion from the steel-producing divisions of the industry, employers retained the internal labor market but altered the administrative procedures. Once employers gained control over the hiring, promoting and dismissal of employees, internal labor markets functioned to intensify job and wage competition in each establishment."[81] Applying the principle of seniority to promotions would have allowed employee representatives to capture some of the power the Amalgamated Association once had in eastern plants but had never exercised in Pueblo. This would have made the ERP a close substitute for the independent union that did not exist for the great majority of steelworkers during the 1920s.

When the company and the employees agreed on a revised plan in 1921, new language championed by the steelworkers in the clause entitled "The Selection and Direction of Working Forces" read: "In making promotions, primary consideration shall be given to length of service and ability to do the work required."[82] The vagueness of this clause explains why employee representatives were forced to fight over seniority throughout the 1920s, but this was still better than no consideration of length of service. If, as one worker complained to Selekman, "the best handshaker gets the job," any recognition of seniority directly threatened the power of the foremen. In fact, as Selekman explained, "Any employe who suspects that he had been discriminated against in a possible promotion has a check on the power of the foremen in the right of appeal, either in person or through his representative under the plan, and in certain circumstances this has been done with satisfaction to the men involved."[83] Employee representatives won many victories on this front, and not just with respect to individual cases. For example, at the March 18, 1921, quarterly conference, in response to a demand from employee representatives, CF&I president Jesse Welborn banned the practice of foremen hiring relatives without the approval of more senior executives in charge of personnel.[84] A strict seniority system might have limited such abuses as well as frivolous appeals, but management refused to give up its prerogative to work outside that system even when pressed. On the other hand, the more often management recognized seniority as a principle when promoting workers, the more likely it became that skilled workers would maintain positions higher on the job ladder at the mill, frequently at the expense of the less-skilled, often Mexican or Mexican American workers with intermittent employment at best.

Numerous sources relating to the ERP at the mill demonstrate the importance of seniority in deliberations between management and employee representatives from the first years of the plan's operation. Undoubtedly in response to pressure from employee representatives, management claimed as early as 1918 that it retained a prerogative to promote the best man for the job rather than the person who had served in the lower-rung capacity for the longest period.[85] Nevertheless, it forfeited that right when it signed a contract with railroad workers on the Colorado and Wyoming Railway. The Brotherhood of Locomotive Firemen and Enginemen and the Brotherhood of Railroad Trainmen convinced management to accept language concerning seniority and promotion in their contract, which included three articles in fifteen sections.[86] "[P]rinciples of seniority to apply in maintaining, reducing and increasing working forces" was also one of the steelworkers' demands to CF&I management before the start of the 1919 steel strike.[87]

During the 1920s, employee representatives at the steel mill registered frequent complaints about management's failure to respect seniority when it came to firing, rehiring (after a layoff resulting from a lack of production), and promotion. As early as November 1921, the secretary of the employee representatives at the plant complained that "a square deal was not given to some of the wage earners at the time of the shutdown last August. He mentioned instances, but gave no names, of wage earners of long service who had been replaced by men from the chemical laboratory and time keeping department who had been in the service but a short time." Management did nothing in response.[88] Perhaps with this setback in mind, Minnequa Works employee representatives offered a four-point resolution for management's consideration at the May 1922 joint conference to rectify a situation they said "may lead to discord or dissatisfaction":

1. That in event of a promotion in that department, the oldest employee in length of service (in that particular line) be given the preference.
2. That in the event the oldest employee in point of service is given a fair trial and is proven incapable the next older employee is to be given the preference.
3. That in event it becomes necessary to reduce the force, that employees will go back in the same order as when advanced.
4. Whenever necessary, employees should be given full opportunity to become familiar with the work next in line in promotion.[89]

In essence, this proposal simply elaborated on the seniority clause in the latest version of the ERP. The representatives wanted to make this provision binding, but management refused, claiming management prerogative.[90] Nevertheless, the demands continued.

None of these roadblocks stopped the steelworks employee representatives from pressing for seniority. In 1924, an employee named Don Williams presented his case for promotion before the Joint Committee on Cooperation, Conciliation, and Wages. The superintendent of the Open Hearth believed Williams was unqualified to serve as first helper. He wanted to select one of the many "better men" available in the department, but none had as much seniority as Williams. The entire committee—the six members who were employee representatives *and* the six members who were part of management—voted to make Williams first helper strictly on the grounds of seniority, despite the fact that "management . . . time and again stated" that strict seniority was "not recognized as being in the best interests of either management or employees in the operation of a steel plant."[91] B. J. Matteson, who reviewed the case for Welborn, explained that this kind of problem could spread beyond the issues. "A public airing of criticism to management is all conducive to contempt of employees toward management," he wrote. He also called this attitude "the direct cause of department management being backward in enforcing discipline."[92]

As a result of such warnings, management eventually asserted itself on this front. In 1927, President Welborn published a letter in a special issue of the *Industrial Bulletin* that contained meeting minutes that read in part, "Length of service is an important element in considering men for promotion in business organizations. It alone, however, cannot be the controlling element."[93] For all intents and purposes, this position demonstrated that the line in the plan regarding seniority was functionally worthless, as management could ignore seniority anytime it saw fit. This issue, therefore, became another reminder to employee representatives of who actually held the power within the ERP.

At the mines, the frustration that accompanied this realization surfaced during the strike led by the Industrial Workers of the World in 1927–1928. Since the steelworkers lacked a viable outside union, they did not have this outlet. Therefore, they simply kept trying to make the most of the organization they had. Yet management's change in attitude toward the plan was more important than what labor thought. Following the 1927 mine strike, management in the mill dropped any pretense that it still believed in the principles John D. Rockefeller Jr. had once championed. This marked the beginning of the end of the Rockefeller Plan.

Chapter Seven

NEW UNION, SAME STRUGGLE

> I am a wobblie and I am glad I am a tool against the Rockerfeller [sic] Plan.
>
> —CF&I MINER SAM CASADOS, IN A LETTER TO HIS UNCLE, JOE BAROS, THAT MANAGEMENT INTERCEPTED, 1927[1]

On the morning of November 21, 1927, five weeks into a strike that had stopped production at most coal mines across Colorado, approximately 500 miners and their wives arrived at the north gate of the Columbine Mine. Permanent replacement workers had kept this mine running during the strike, so deputy sheriffs and state police were there to meet the group. "Who are your leaders?" the head of the state police contingent shouted to the strikers. "We are all leaders!" they responded in unison. After more taunts were exchanged, strike leader Adam Bell asked the police to unlock the gates to the mine. When he put his hand on the gate, a policeman struck him with a club. Other police started to throw rocks at the miners, and the miners threw them back at the police. When the miners forced their way through the gate, the police fired into the crowd, killing 6 and wounding at least 25.[2] This incident became known as the Columbine Massacre.[3] It was the deadliest but by no means the only violent incident to occur during the Colorado mine strike of 1927–1928.

Reaction to this tragedy by the Industrial Workers of the World (IWW), the radical union behind the strike, was swift and severe. Speaking to reporters in Denver, a representative of the Wobblies blamed the killings on the "hired assassins of John D. Rockefeller Jr." As was the case in 1914, a crowd of socialists appeared outside Rockefeller's office the next day carrying signs bearing slogans such as "John D. Rockefeller. Your Gunmen Will Not Break the Solidarity of Labor."[4] The problem with this suggestion was that John D. Rockefeller Jr.'s Colorado Fuel and Iron Company (CF&I) did not own the Columbine Mine; it was owned by the Rocky Mountain Fuel Company. Josephine Roche, who had just inherited the business from her recently deceased father, insisted that the strikers' rights be respected and had agreed to let them picket freely in the nearby town of Serene as long as they did not interrupt the operation of the mine.[5] Even though CF&I did not own the Columbine Mine, the struggle for unionization was very much the same as it had been back in 1914, and many of the soldiers who had been at Ludlow also took part in the Columbine Massacre.[6] Therefore, it is easy to see how a public that remembered the Ludlow Massacre might not necessarily know the difference between the two companies.

The IWW's attempt to connect the Columbine Massacre to Rockefeller demonstrates its perception that twelve years of the Rockefeller Plan had done little to improve the industrialist's reputation among the public, not to mention among many Colorado coal miners. It also explains why the IWW and its sympathizers often invoked the spirit of Ludlow in 1927–1928 to motivate those involved in their struggle. "Remember the massacre at Ludlow!" wrote Max Shachtman in *The Labor Defender*. "The movement of solidarity must be swiftly built."[7] Bringing Rockefeller into this dispute not only attracted attention but also increased support for the strike. In the striking mines actually owned by the Colorado Fuel and Iron Company, the walkout became a referendum on the employee representation plan (ERP). From labor's standpoint, twelve years of the Rockefeller Plan meant nothing. The plan's existence did nothing to stop Colorado's Great Coalfield War from resuming.

ORGANIZING AGAINST THE ROCKEFELLER PLAN

In April 1924, CF&I management used the Rockefeller Plan as a vehicle for more wage cuts. As was the case in 1921, management "asked" employee representatives to approve a wage cut in the name of the miners they supposedly represented, which they did. A second cut implemented the same way came in March 1925. Taken together, the two decreases of 20 percent each returned wages to 1917 levels. The first resolution the employee representatives signed off on explained that the cuts were supposed to provide more steady employment for everybody in the mines. Such a "voluntary reduction in wages," read the resolution, was therefore "for the best interest of the employes."[8] This apparent stamp of approval made it more likely that the Colorado Industrial Commission would approve the action.[9] However, as with many employee-initiated actions undertaken under the auspices of the Rockefeller Plan, not all workers signed on to these arrangements voluntarily. At one mine, management fired sixteen miners who refused to sign the 1925 petition.[10] When every miner at Berwind and Cameron refused to sign, management closed those mines.[11]

Despite claims to the contrary, management understood that many miners opposed these reductions. The CF&I Archives has a copy of a petition to the Colorado Industrial Commission complaining about the cuts for such reasons as "aforesaid mine has not worked more than half time during the last [t]wo years" and "in the living conditions of 1917 and 1925 there is a vast difference." A total of 114 miners signed the petition.[12] A meeting at Coal Creek attended by 300 miners from Coal Creek, Rockvale, and Fremont passed a motion calling on the Colorado Industrial Commission to rescind the decrease. Yet the CF&I executive in charge of mining, David Stout, responded, "We feel quite sure, from the expression of our employees, that they wish to be let alone and the agreement which they have entered into, is entirely satisfactory to them."[13] It could not have helped management that miners in the United Mine Workers of America (UMWA) had just been awarded $7.75 per day under a contract known as the Jacksonville Agreement (after the town in Illinois where the agreement was signed). CF&I miners doing shift work (the best-paid miners) made $6.20 a day after the cuts. To make matters worse, management cut wages again before the 1927 strike began.[14] The use of the ERP in this manner, particularly after management had used it the same way in 1921, infuriated miners of all stripes, whether they favored the IWW, the UMWA, or no union at all.

In response, the IWW made its critique of the Rockefeller Plan the centerpiece of its organizing drive. As a September 15, 1927, open letter to John D. Rockefeller Jr. explained:

For five years your scheme of Industrial Democracy has functioned without interference from any of the Labor Unions in your principality of coal in Colorado.

As a result of that we, miners, of Colorado find that our wages have been cut more than $2.00 a day and more when we are working on contract work. We were promised that prosperity would follow with Industrial Peace. When our wages were cut we were told that we should have more steady work, but the fact today is, that we work less days a year than when we had industrial strife.[15]

The strategy worked. As former Wobbly E. W. Latchem later wrote, the ERP "was the real 'sore-spot' which made possible IWW success in 1927."[16] Ronald L. McMahan, who along with Eric Margolis interviewed many retired miners as part of an oral history project at the University of Colorado in the mid-1970s, concluded, "It was to a large degree the gross inadequacies and failures of the Rockefeller plan, not simply the cut-throat competitive conditions that typically surrounded laissez-faire capitalism, that set the stage for the 1927 strike."[17] In short, membership in the IWW in 1927 constituted an explicit rejection of John D. Rockefeller Jr.'s version of industrial democracy.

The Industrial Workers of the World had formed at a famous meeting in Chicago in 1905. Government repression of the group because of its opposition to World War I had effectively driven the IWW underground during that conflict. Nevertheless, the IWW started organizing in Colorado coal mines in 1919, when a few organizers arrived with literature to distribute at poorly attended meetings.[18] In 1922 one coal mine manager reported that he "had met several I.W.W. agents and firebrands during the last few days" and that they were "very active in getting located." Two Wobblies had been given jobs at the Segundo Mine and had joined the UMWA to proselytize.[19] The organizing accelerated with the arrival of IWW leaders Frank Jurich in 1925 and A. S. Embree in March 1926. The union centered its organizing efforts in Walsenburg, the center of CF&I's southern Colorado properties, holding its first mass meetings there in March and July 1927.[20]

Yet the strike was triggered almost by accident. The IWW's international executive board called upon all union members to stop working for three days starting August 8 to protest the execution of Massachusetts anarchists Nicolo Sacco and Bartolomeo Vanzetti. This was a nationwide effort, not specifically directed toward Colorado. However, miners in Colorado responded particularly well to the call. In fact, almost all CF&I miners honored the shutdown.[21] Furthermore, unlike other strikes in honor of Sacco and Vanzetti across the country, the one in southern Colorado moved into a second day. Only when the Colorado Industrial Commission condemned the action did most miners return to work on the third day.[22] With such results for a mere sympathy strike, IWW leaders realized for the first time just how much support they had in Colorado's coal fields.

Following this successful practice run, IWW organizers spread throughout the state in anticipation of another upcoming action. On September 4, 1927, 187 delegates from around Colorado met at Aguilar to set a strike date and draw up demands that included a six-hour workday, a five-day week, and a substantial raise. They set the strike to begin on October 8. In response, CF&I submitted a proposal to its employee representatives that raised wages from $5.52 to $6.20 per day effective October 1. The IWW wanted $7.50, the rate stipulated in the Jacksonville Agreement.[23] The IWW did not bargain in the traditional sense, so its demands did not include union recognition. However, the miners were willing to send representatives elected by union sympathizers to meet with the coal companies—a backhanded slap at the legitimacy of the Rockefeller Plan.[24] The convention delegates also included a requirement that IWW organizers be allowed free access to all mining camps, something the Rockefeller Plan supposedly already ensured.[25]

The Wobblies' demands did not sit well with the authorities. Even though the miners had given thirty days' notice of their pending labor action, as required by Colorado law, the Colorado Industrial Commission still ruled their strike illegal because it did not believe the delegates in Aguilar really represented the miners. In fact, because the Colorado Industrial Commission considered the IWW an outlaw organization, it would not have respected the results of any meeting the union organized.[26] An "outlaw organization" could only organize outside the protection of the law. Since CF&I management had already begun to violate the union's nondiscrimination clause in the Rockefeller Plan after the 1919 strike, disaffected miners could not count on that arrangement for protection.

THE STRIKE BEGINS, THE EMPLOYEE REPRESENTATIVES REACT

After the Colorado Industrial Commission declared the strike illegal, IWW leadership moved its start date from October 8 to October 18.[27] Felix Pogliano, secretary-treasurer of UMWA District 15, predicted that the strike would never occur.[28] Yet in light of the successful Sacco and Vanzetti walkout two months earlier, the actual response should not have surprised anyone. As Frank L. Palmer explained in *The Nation* a few months later, "Before the strike was called I.W.W. organizers bragged that 4,000 miners would walk out. Actually, 5,500 struck. And the strike spread."[29] It hit both northern and southern Colorado operators hard. In Huerfano County, where CF&I was the dominant player, only 1,180 of 2,594 miners showed up for work. In Las Animas County, another CF&I-dominated area, only 500 of 3,400 miners crossed the picket line. Interestingly, Fremont County (in southern Colorado, west of Pueblo) proved the exception to the rule; miners came to work almost in full force that first day. However, this impressive turnout was a result of the strong presence of the UMWA, which refused to follow the IWW strike call.[30] Every CF&I mine continued to operate despite the strike, although production decreased significantly for obvious reasons.[31]

Not every miner who participated in the strike necessarily agreed with the IWW's radical agenda. As Merle Vincent of the Rocky Mountain Fuel Company later explained to the Colorado Industrial Commission:

> In the dispute immediately confronting us the miners charge that conditions are the cause of the strike. Operators have denied this assertion and charge that the I.W.W. is the cause of the strike. An impartial observer with accurate knowledge of facts and circumstances attending the strike would probably declare that a number of causes contributed to and resulted in the strike in the Colorado coal fields; that while this organization is immediately responsible for the strike call, conditions are primarily the cause of some 5,000 miners striking.[32]

The best evidence for this assessment comes from the platform for the strike. Adopted by the 187 delegates at the meeting in Aguilar on September 4, it included higher wages, an end to work on Saturdays and Sundays, elected check weighmen, payment for all necessary "dead work" such as timbering to prevent tunnels from collapsing, and a cap on rent and utility costs for company-owned housing.[33] These demands offered no hint that they had been drawn up by an incipient revolutionary organization since that might have alienated moderate miners whose support the IWW needed. The fed-up miners who made these demands were willing to give any union willing to organize them a chance to fight for what they wanted. With the UMWA not available to most Colorado miners, the workers turned to the closest union at hand.

The decision of distinctly nonradical workers to turn to the IWW again tested the limits of management-imposed democracy. While the clause in the employee representation plan that referred to "no discrimination" on the basis of "membership or non-membership" in a union made no exception for radical unions, CF&I management could not tolerate the thought of Wobblies on the company's payroll. In the past, company executives had talked and bargained with miners who were both employee representatives and UMWA members, but the day after Labor Day "the Company discharged without notice . . . eight employees who had made it known in one way or another that they were affiliated with, or had been active in spreading propaganda for, the I.W.W." However, Kenneth Chorley, the Rockefeller aide who reported on this firing to the New York office, also explained that management had not started taking action against Wobblies until the Sacco and Vanzetti walkout began in August.[34] Either management was too overconfident to believe the IWW would prove preferable to the ERP, or the miners organized with remarkable discretion.

If management's statements after the strike started can be believed, it seems that IWW organizers went largely undetected before August 1927. One reason may have been management's decision to stop using spies to gather intelligence from employees.[35] Without the spies, management had to rely on employee representatives for intelligence, and they failed to give adequate warning about the strike risk the company faced. As CF&I president Jesse Welborn testified before the Colorado Industrial Commission near the official end of the walkout, "[M]any meetings with the employes of the Colorado Fuel and Iron company were held prior to October 18th at which the employes expressed themselves as generally speaking satisfied with conditions and they further expressed the feeling that any matters that might, in their opinion, need adjustment, could be easily handled through the regular procedure under the plan."[36] While Welborn may have been less than candid with the commission, his surprise at the events makes sense. By 1927, the miners who served as employee representatives were the ones least likely to strike. Most were native-born, white, and had long tenures with the company.

Even though the representatives did not strike in this instance, they still had an agenda to press before management. During the strike, employee representatives sought better wages through their union. Asked during an October 24 meeting for suggestions as to how to get the miners back to work, Adam Young of Lester replied, "Only suggestion I have to offer is that you raise the wages back to $7.75, and the miners the same as they were getting then. They will then all go back to work."[37] Management responded that it had already increased wages as much as it could.[38] Nevertheless, the employee representatives continued to press management on this issue. Management held another meeting in the Walsenburg District on December 28. As President Welborn described it the next day, "The Representatives presented further arguments for increases over those granted October 1st. . . . [D]esiring as nearly as possible to meet the wishes of its employees, and with the hope the action would receive general approval, the officials agreed with the Representatives to adopt, effective January 1, 1928 [the] . . . basic day rate $6.52 in all districts."[39] When the strike ended, the Wobblies claimed credit for winning this wage increase from management. While this claim has some validity, the employee representatives pushing for the same demand from the inside undoubtedly helped make it happen as well.

CF&I workers' reaction to the strike reflected the racial and ethnic divide between the employee representatives and rank-and-file miners. The Industrial Workers of the World took a stand against racial discrimination from its inception in 1905. The early history of the Wobblies, including the organization of lumber workers in Louisiana and of copper miners in Arizona, demonstrated the union's adherence to this principle. So did this strike. "So far as we can learn," Welborn explained to two Rockefeller aides in a letter on September 14, "the membership obtained by the I.W.W. leaders is very largely composed of Mexicans, most of whom are comparatively recent immigrants and are probably still citizens of Mexico. The native Mexicans [Mexicans who were U.S. natives], who represent a substantial percentage of our coal miner employes, are, generally speaking not interested in this movement."⁴⁰ It was only logical that Mexican immigrants and Mexican Americans would form the core of support for the walkout because they had the worst jobs and significantly less representation on the ERP than did white and native-born workers. In fact, the IWW was so confident of its support among minorities that its main organizer claimed it did not even need native-born Americans.⁴¹

With a racial composition skewed toward nonwhites, the IWW inevitably came under attack not just for its politics but also for its supporters' racial and ethnic backgrounds. For example, *The Rocky Mountain News* blasted the strikers as coming "from a land of revolution and counter-counter revolution. . . . [M]any have been taught to know no law but the law of might. A migration over the border does not change that point of view. Being in a majority in [a] number of Colorado mines they are not inclined to accept American ways."⁴² In an executive proclamation issued after the strike began, Colorado governor William H. Adams called the IWW "un-American," a clear double meaning.⁴³ It is therefore no surprise that the strikers often carried the American flag during their marches. Indeed, the strikers who marched on the Columbine Mine carried a banner that read "We Are Americans."⁴⁴ Such tactics resemble the rhetorical creation of the racial group "Spanish American" to show that some people of Mexican descent deserve to be considered white. However, the hostile response to the strike by management and its sympathizers demonstrated that people with economic power played the biggest role in determining who belonged to what racial categories.

Like its sympathizers in the press and in government, CF&I management showed a new willingness to explicitly divide workers by race during the 1927 strike. For example, the CF&I Archives includes the transcript of a conversation with miner John Shepherd dictated by Walsen Mine manager J. L. McBrayer following the Sacco and Vanzetti walkout. It reveals that the company brought racial divisions into its anti-strike efforts at a personal level:

> McBrayer: "I called Shepherd over, and said, 'John, I am surprised at you. You are a white man; Scotch, same as I am; you come of a good family. . . . Haven't I always treated you well?'"
>
> Shepherd: "Yes, you have always given me a square deal ever since you have been at Pictou."

McBrayer: "Then what is wrong? You know you are with a bunch of outlaws. Don't you realize you have ruined your reputation in this Country? . . ."

Shepherd: "Well, wages and working conditions are not right."

Further pressed by McBrayer, Shepherd admitted that he had never taken a grievance to an employee representative despite the fact that he indeed had many.[45]

Why did management appeal to racial divisions now when it had once tried to champion civil rights? The increased number of Mexicans and Mexican Americans in the industry, along with the tendency of Mexicans to go on strike, explains why. If management had tried dividing workers by race earlier, there would have been too few Mexicans to create significant divisions, but there were more than enough to do so by 1927.

If, as Rockefeller and King often suggested, management and labor truly had overriding mutual interests, management should have been able to appeal to those interests rather than resort to the race card. The 1927 strike, therefore, marked a significant turning point in the history of the plan. While management had at least returned to giving lip service to the principle of equality following previous strikes, once it began to bow to and even encourage racial prejudice among native-born employees, it probably recognized that it was permanently alienating minority workers from participating in the plan. Ironically, at a time when trade unions in Colorado were arguing that Rockefeller still directed everything, such a change was possible only because he had effectively delegated oversight of industrial relations at CF&I to others. This allowed his staff to implement significant changes to his original vision for the plan after the strike ended.[46]

MINNEQUA WORKS REPRESENTATIVES' MEETINGS AND THE COLORADO INDUSTRIAL COMMISSION HEARINGS

Governor Adams used the Columbine Massacre as an excuse to send the National Guard to mines in both southern and northern Colorado in November 1927. With guards in place for protection, miners gradually returned to work underground. Jesse Welborn again blamed fear of violence as the reason his miners did not report to work. Yet even though the northern Colorado mines reopened as a result of the presence of more law enforcement personnel, in the south the strike still smoldered. In response, CF&I management continued to use employee representatives to make its case to both miners and the public. In September, Andrew Diamond introduced a resolution at the quarterly meeting of the steelworks ERP requesting that management fire all employees there who were discovered to be IWW members.[47] In November, Diamond and a group of other steelworks representatives toured CF&I's southern Colorado mines, meeting with employee representatives and asking what they could do to put an end to the strike. In one sense, this course of action was entirely understandable because the coal strike had curtailed operations at the Minnequa Works. A lack of coal to fuel the blast furnaces had put 2,500 men out of work. Diamond promised the miners' representatives at Ideal that "the officials of this company have nothing to do with this meeting. Anything that transpires here is just like it was in our own monthly meeting at Pueblo."[48] Nevertheless, the transcript of this and every other meeting Diamond had with the miners ended up in the company's files.

By December 1927 the IWW had put all its hopes for resolution into hearings of the Colorado Industrial Commission designed to force the enactment of a pay increase, the prime motivator behind the strike. To head off the need for such action, management instituted a wage increase that month in response to another employee representative request. Kenneth Chorley reported that the company had first used this tactic back in September:

> At our lunch on September twentieth Mr. Welborn and I both came to the conclusion that it was necessary for the Company to take some quick action in increasing the wages of the miners. I suggested to Mr. Welborn that it seemed to me here was an opportunity to deliver a strong blow at both the I.W.W. and the United Miners if it could be worked so that a request for increased wages could come through the regular channels of the Employee Representatives Plan, that the company would simply take the position that they would have nothing to do with the outside labor organization, that they had well-regulated machinery set up for such procedure and that when a request was presented through regular channels they would be only too glad to negotiate and work out whatever seemed to be right. Mr. Welborn, while appreciating the benefit of such a scheme was somewhat fearful of going against the precedent of having all wage increases made by the Company.[49]

The novelty in this situation came from using the ERP to raise wages rather than lower them. Welborn feared one employee representative wage increase request would lead to many others, but because of the strike he twice raised wages anyway. While this second wage concession did help to bring many more CF&I miners back to work, it did not eliminate the need for hearings.

The Colorado Industrial Commission began its hearings into the strike on December 19, 1927. The first hearings dealt exclusively with northern Colorado firms but still proved embarrassing to the coal operators' cause. Miners representing all Colorado coal companies, including Mexican nationals speaking through interpreters, brought scores of complaints before the commission. The commission did not come to southern Colorado to take testimony until January 3, 1928. It held hearings in Colorado Springs, Cañon City, Walsenburg, and Trinidad.[50] Notes in the CF&I Archives regarding that testimony include complaints about the failure to obtain timbers so miners could prop their tunnels and dig safely; unsafe electrical wiring in the tunnels; a shortage of cars to bring the coal the miners dug aboveground; inability to select their own check weighmen; mine closures that left them out of work for months at a time; slowness of the man trip (which hoisted miners out of the mine), delaying the miners from leaving on time each day; and favoritism with regard to promotions to contract work that offered better wages. Such grievances should have been addressed by the Rockefeller Plan. This testimony served as incontrovertible evidence that the ERP had broken down. When the commission asked the miners why they either left CF&I or joined the strike, the invariable answer was that they "could not make [a] living" from the wages the company paid.[51]

Other testimony took direct aim at the Rockefeller Plan. As miner Conrad Avillar, a member of the IWW committee directing the strike, explained to the commission, "They tells us about representatives of the Rockefeller Plan—they are in the same position as the miners. They can't do anything else for the miners because they are afraid they will lose their jobs or employment. Not only they don't do anything for the miners, but I have seen time and again when they go and testify against the miners for fear of losing their own jobs, I suppose."[52] Joe Vosnica, who worked at the Robinson Mine, made a similar argument during his testimony:

Q. Did you belong to the Company union at the mine?
A. Well I suppose you mean under the Rockefeller plan, but I never believed in it.

Q. When you had difficulty in getting your track material and those other delays you have spoken about, did you at any time take the matter up with the Company's representative?
A. No, I did not.

Q. Are you able to tell the Commission why you did not do that?
A. Well, just because it did not pay, it did not make any difference; that representative has just as much power as I have, and if I told him he would tell me the same thing, that it was in Pueblo, or go and find them or something. He did not have any business whatever to take it up with the boss or superintendent, and if he did he would tell me the same thing.

In the margins of one copy of Vosnica's testimony held in the CF&I Archives, someone from management scribbled, "Representative did have power to take up with boss or [superintendent]. If brought to his attention."[53] While technically true, the fact that miners like Vosnica had lost faith in the plan made that point irrelevant.

After January 1, more and more miners returned to work to earn the money they needed to survive. On February 19, 1928, 88 percent of the miners who voted chose to return to work the next day, pending the release of the Colorado Industrial Commission's report. Nevertheless, the IWW claimed partial victory, based largely on the potential for future action by the organization created as a result of the walkout. "True, you have not won all your demands," wrote A. S. Embree in a strike bulletin. "But did you ever expect to win them all?"[54]

The Colorado Industrial Commission released its final report on the strike on March 20. The commissioners, like much of the general public in Colorado, had nothing nice to say about the IWW. "The Industrial Workers of the World," they wrote, "is the advocate of principles that, in our opinion, should they be successful, would mean the destruction of our system of government." Later in the report, the commissioners echoed the criticism management had made against the IWW during the strike: "The representatives of this organization came to Colorado from other states for the purpose of creating trouble in the coal mining industry. The majority of these men are professional agitators who have never worked in a coal mine." They also attacked the IWW for its failure to offer constructive suggestions. "It was plainly evident that a strike was what this organization wanted," the commissioners explained. "[N]o attempt was made by the Industrial Workers of the World to adjust the grievances or the wrongs of the miners in a lawful manner. There was nothing constructive in their program. They offered no practical solution for the trouble."[55] For a commission charged with stopping labor trouble before it started, this aspect of the Wobblies' agenda must have been particularly galling.

Yet despite denouncing the IWW as an organization, the commission still essentially sided with the striking miners by harshly criticizing the coal operators for creating the conditions that had allowed the strike to occur in the first place. Most important, the commission blamed the IWW's success in organizing miners on the "[l]ack of an organization of their choice among the miners." Of particular importance here, the commission implicitly withdrew its previous approval of the Rockefeller Plan. "Very little freedom of contract can exist between a man who has nothing but his labor to sell and the employer who can do without the labor of any particular individual," explained the report. "Freedom of contract under such conditions is little more than an idle dream, because of the self-evident inequality."[56] This is precisely the realization that eluded John D. Rockefeller Jr. and Mackenzie King. The fact that a Colorado government commission could see what they could not is not so much a slight of Rockefeller and King's judgment but rather a sign that labor relations had changed in the years since the plan was introduced. As the commission explained, "The time is past when any employer or owner of a mine can operate it as they please."[57] In many ways, Franklin D. Roosevelt's labor relations policy during the New Deal would be built on this same premise.

Although the miners still had no union when the strike ended, the commission's report was at least a great moral victory for them, if not for the IWW. The report inspired the Wobblies in Colorado to keep pressure on the coal companies even after the strike ended. "Now is the time for action," explained a Wobbly newsletter on April 20, 1928. "Every job delegate, and in fact every member should make his business to see that the man whom he is working along side of has an I.W.W. card."[58] The union's leadership hoped to start another strike by the end of the year. The miners had won important concessions on wages and working conditions because of the strike, and they hoped to win even more. As it turned out, however, they would do so through a different union than the IWW.

THE IWW VS. THE UMWA

The United Mine Workers of America and the Industrial Workers of the World had little regard for each other. In fact, the UMWA had banned its members from carrying IWW cards under threat of expulsion back in 1919.[59] "The I.W.W. seeks to destroy Trade Unions," explained a statement from UMWA District 15 issued at the height of the 1927–1928 strike. "It is the agent of destruction. The very essence of its being is revolutionary, therefore, we cannot subscribe to such doctrines."[60] For its part, the IWW had denounced the UMWA as "a voice from the past" as early as 1923.[61] The UMWA was practically dead in Colorado after its failed 1922 strike. "As you are doubtless aware," wrote Colorado UMWA president Frank Hefferly to John L. Lewis in late 1924, "the organization in District Fifteen has dwindled away until it might safely be said that even the district organization, to say nothing of the local unions, has practically ceased to function."[62] In April 1927, eastern and midwestern coal companies locked out their unionized miners, a situation that threatened the union's very existence.[63] Under these circumstances, the UMWA could not have mounted a successful organizing campaign in Colorado in 1927–1928. While the UMWA agreed that the issues the Wobblies wanted to change deserved to be adjusted, it refused to support the walkout because of the politics of the union that organized it.[64] Therefore, it did nothing. Ironically, the UMWA nonetheless ended up as the chief beneficiary of the IWW's efforts.

The meme that gave rise to the UMWA's triumph began during the strike when the press (along with the Colorado Federation of Labor and later the Colorado Industrial Commission) started to blame the failure of the Rockefeller Plan on the rise of the Industrial Workers of the World.[65] As *The St. Louis Post Dispatch* editorialized:

> In the West, there exist bad conditions and a limited work force. Having no alternative, the miner opts for the I.W.W. which offers him a chance to use a method of industrial warfare which constructive

unionism frowns upon. The miner, of course, is in error in allowing himself to be thus misled, and he will no doubt pay dearly for his mistake. The false philosophy and destructive tactics of the IWW have no place in our industrial order. But it is well to remember that if some of the mine owners had used wiser policy with regard to unionism, the western miner would never have erred.[66]

Local newspapers throughout Colorado looked specifically to Rockefeller to end the walkout. *The Denver Evening News* was particularly blunt. "Has Mr. Rockefeller wearied of welldoing?" it wrote. "The newspapers believe that if the unfortunate conditions now existing in Colorado mining districts were put squarely before Mr. Rockefeller he would become sufficiently interested to take a hand in the business and bring about a settlement with the legitimate coal miners of the state."[67] Like the IWW, they remembered the industrialist's promises during his 1915 and 1918 tours of the state, and like the IWW they called on him to live up to those promises.

This time John D. Rockefeller Jr. had no new angle on industrial relations. The fresh perspective needed came from Josephine Roche, the new president of CF&I's much smaller rival, the Rocky Mountain Fuel Company. On August 16, 1928, her company signed a contract with the United Mine Workers of America, even though, as Roche later put it, the union had "no power locally to enforce its requests and was losing ground nationally" at the time.[68] The choice of that union was no accident. A few months earlier, the company had announced its willingness to make a contract "whenever the miners are organized in a union affiliated with the American Federation of Labor."[69] The contract with the UMWA, which included the shared goal "to stabilize employment, production, and markets through cooperative endeavor," served as protection for Rocky Mountain Fuel from the instability the Wobblies cultivated for their revolutionary ends.[70] Recognizing this reality, the IWW charged, "The U.M.W.A. has become a COMPANY UNION. It disorganizes instead of organizing the workers. Its officials are henchmen of the employers and traitors to the workers."[71] The Wobblies made this allegation despite the fact that their leadership had considered signing its own contract with Rocky Mountain Fuel while the company was still willing to talk to the union.

The presence of the IWW in Colorado made the UMWA seem conservative and safe by comparison. One of the principles outlined at the beginning of the first contract between Rocky Mountain Fuel and the UMWA suggests that the two parties had the IWW in mind when they formed their agreement. It read that the parties intended "[t]o defend our joint undertaking against every conspiracy or vicious practice which seeks to destroy it and in all other respects to enlist public confidence and support by safeguarding the public interest."[72] A CF&I labor spy report in late February 1928 suggested that the IWW had begun planning another strike, which was likely the reason Roche wanted to recognize the AFL so it could defend Rocky Mountain Fuel against the IWW.[73] "Taking everything into consideration, the southern district from an organization standpoint is going along fine," reported the IWW newsletter the previous April. "Job delegates are all in the camps and they are doing their work well. Although the job conditions are rather poor yet many of are [sic] best men are back in the mines."[74] By joining forces with the UMWA, Josephine Roche cut off the IWW organizing effort and encouraged the UMWA to return to the Colorado coal fields to start organizing again in earnest.

No other Colorado coal company willingly signed a contract with the UMWA until 1933, despite offers from the union to sign contracts comparable to the one in place at Rocky Mountain Fuel. In fact, CF&I refused to continue to distribute Rocky Mountain Fuel's lignite coal (with which CF&I's coal did not compete) through its sales network after Roche signed the union contract and also refused to buy Rocky Mountain Fuel coal on the wholesale market.[75] Nevertheless, CF&I and other Rocky Mountain Fuel Company competitors might have benefited from this new direction in industrial relations, as the union agreement brought peace to Roche's struggling company. A few weeks after the agreement had been reached, *The New York Times* reported, "It would seem that in the ordinary course of events the deepest hostility would prevail between the miners and the owners of the Rocky Mountain Fuel Company. Strangely enough, there exists at the present time the most cordial relations between the company and its employees."[76] The correspondent logically attributed this goodwill to the UMWA contract, but it was more than that. One sign that Roche actually meant what she said was her decision to hire the surviving union martyr of the Ludlow Massacre, John R. Lawson, as the company's vice president.[77] The goodwill generated by such actions was so great that the Colorado Federation of Labor publicly endorsed Rocky Mountain Fuel

Company coal to the buying public.[78] In fact, the UMWA invested union funds in the company.[79] The goodwill continued into the Depression years, as the miners agreed to postpone receiving half their wages during the fall of 1931.[80] The situation was much different in the mines of the Colorado Fuel and Iron Company.

THE END OF THE ROCKEFELLER PLAN IN THE MINES

During the strike (and no doubt because of it), CF&I management willingly admitted for the first time that it might not be hearing all of its miners' complaints. A notice posted on January 10, 1928, and signed by everyone from the local employee representative to the general superintendent of the company's mines read:

> We occasionally hear that coal mine employes have grievances that they do not express to Mine Foremen or others for fear of discrimination or discharge.
>
> Our Employee Representation Plan guarantees every employe the right to present grievances to those in authority without danger of discrimination.
>
> It is our aim to have employes satisfied with their working and living conditions and we . . . take this means of assuring the employes working under our direction that we desire that they inform us about any condition that is unsatisfactory to them. All complaints will be carefully investigated and no one will be discriminated against for complaining about that he thinks is wrong.[81]

This was not the only change to the plan or the company following the start of the strike.

In 1929, Jesse Welborn retired from his post as CF&I president to assume a position as a director of the company. He was replaced by Arthur Roeder. Roeder had been an executive at the American Linseed Company, a firm the Rockefellers had invested in heavily.[82] The Depression would be Roeder's trial by fire, but it hurt CF&I employees even more. As Welborn, serving as chair of the Citizens Employment Committee in Denver, explained in 1932, "With the new cases that apply daily for work . . . are many old employees of the C.F.&I. Co. These include clerks as well as men from our mine and steel operations."[83] Relief operations for coal miners and their families began in the fall of 1930. Even though management oversaw the relief efforts at every mining camp, employee representatives served on subcommittees that determined how those efforts at each camp were carried out. As part of these activities, management handed out surplus clothing it bought in bulk, distributed seeds so miners' families could start vegetable gardens to feed themselves, and trained miners to pan for gold to replace lost income from the mines. These measures were not enough. Along with these relief efforts, the company extended credit at its company stores to the extent it felt it could and suspended rent payments on company houses for under- and unemployed miners.

Yet even these well-intentioned efforts became another reason for some employees to hate the Rockefeller Plan. "In fact," wrote G. H. Rupp, the executive in charge of miners' relief, to management in Pueblo, "it has been necessary to make the representatives more liberal, as their responsibility without previous experience makes even an honest radical rather hard-boiled. A man's first duty is to provide for himself and his family. No employer dictates how a man shall spend his money. This function has been taken over to a great extent for the period of the depression."[84] With this dynamic in mind, it is easy to imagine how these relief efforts might have contributed to, rather than hindered, the success of the United Mine Workers of America's 1933 organizing campaign.

While white miners had serious problems during the Depression, the situation for Mexican and Mexican American workers was even more dire. At Valdez, so many of these workers lost their jobs that CF&I started a special program for them: rug weaving. "Instructors were hired from New Mexico," Rupp explained, "where the well-known Chimayo rugs are made. The Company furnished the material and has endeavored to sell the products. After the course of instruction, home-made looms were built in the various homes and the weaving done there. There are thirty looms now in the homes, and about the same number of really good weavers. A man working steadily weaving makes from $3.50 to $4.00 a day at the present price of rugs."[85] The company helped 200 Mexican nationals who had been laid off at the mines repatriate to Mexico in 1932.[86] According to historian Zaragosa Vargas, Colorado relief agencies, aided by Mexican consulates, repatriated approximately 20,000 Mexicans between 1930 and 1935. Many such efforts were not voluntary and a number of Mexican Americans, especially Mexican American children, were caught up in them.[87] There is no way to tell whether some of these people were former miners whom CF&I helped move, but they might have been.

To make matters worse, CF&I drastically cut wages for those who were able to work during these trying times. On June 1, 1931, company president Roeder wrote to the Colorado Industrial Commission, stating the firm's opposition to any further wage cuts for its miners.[88] Nevertheless, later that same month, coal miners at Crested Butte complained to the commission that CF&I had in fact already cut their wages. Management withdrew the cuts immediately after they were exposed publicly. On July 31, however, CF&I announced a broad reduction in daily wage rates from $6.52 to $5.25 that took effect without immediate complaint. Other firms cut wages at the same time, but as the largest coal company in the state, CF&I drew all the attention.[89] In a petition to the Colorado Industrial Commission, a group of miners at Valdez explained how management enacted the cut: "About four weeks ago or more the miners representatives called a meeting for a slash in wages. Mr. Sena[,] one of the representatives[,] said that we should just as well take the cut as we were going to get it just the same. But the miners refused this so the meeting was adjourned."[90] So while CF&I was still using its old tactics, they did not work as well in the desperate conditions created by the Depression.

The Colorado Industrial Commission's ruling on these wage cuts showed that the circumstances faced by the Rockefeller Plan had changed drastically since the Depression began. The commission explicitly rejected the Rockefeller Plan as a legitimate vehicle for collective bargaining.[91] The wage cuts also drew the explicit ire of Rocky Mountain Fuel's Josephine Roche. These concerns were in part self-interested. If CF&I had significantly lower labor costs than Rocky Mountain Fuel, it could severely undercut her union-dug coal in the marketplace. Nevertheless, Rockefeller's past public statements allowed Roche to make the concerns she expressed in a telegram to him seem entirely high-minded:

> One word from you can prevent a recurrence of the human and economic waste which will result from the action taken by your company, the Colorado Fuel and Iron Company, in cutting miners' wages twenty per cent. For forty years industrial conflict has periodically broken out in Colorado as a result of similar attempts to secure operating profits at the sole expense of the workers. Following the Colorado loss of life

and property in nineteen fourteen which culminated in the Ludlow massacre you widely advertised a new industrial program by public assurances which now take on fresh importance, to the effect that conditions leading to industrial upheaval in Colorado would never recur. But the causes of industrial unrest were not removed, and the traditional and anti-social methods of the past are again being employed by your company.[92]

Matthew Woll, vice president of the American Federation of Labor, made a similar public personal plea.[93] Rockefeller offered no response to either of these statements. To make matters worse, CF&I tried to cut wages again in 1932, shortly after the commission stopped an attempted cut by its non-union rivals. A total of 695 CF&I employees signed a petition protesting the action. The commission emphatically rejected this additional 15 percent wage cut as a step too far.[94]

It took a new presidential administration in Washington and the drastically new attitude toward organized labor that accompanied it to force CF&I to finally change its policies.[95] Colorado coal miners joined the UMWA in large numbers even before the National Industrial Recovery Act (which offered the first broad federally sanctioned right to organize in U.S. history) passed the U.S. Congress in 1933. American Federation of Labor sources claimed that 95 percent of Colorado coal miners joined the UMWA in a thirty-day period.[96] When CF&I joined other southern Colorado coal operators in agreeing to a union contract that November, 877 miners voted for the contract and 273 voted in favor of the ERP.[97] Even at the end, Rockefeller's vision still remained popular with a significant minority of CF&I miners. Only frustration with the Rockefeller Plan, however, can explain the blinding speed with which the UMWA organizing campaign succeeded. It only took from June (the start of the drive) to October 1933 to gain a union contract.[98] In short, as soon as CF&I miners had a viable alternative to the plan other than the IWW, they flocked to the independent union even though they had no foreknowledge that the company would give in so easily, especially during a depression.

Despite the end of the plan in the mines, CF&I remained reluctant to give up on John D. Rockefeller Jr.'s experiment. "Recent press dispatches state that the Colorado Fuel and Iron Company is abandoning at its coal mines its industrial representation plan," President Roeder wrote shortly after signing management's first agreement with the UMWA. "This is not true. The exceptionally satisfactory relationship between men and management that has existed since the adoption of the plan still continues."[99] While Roeder did not clarify exactly what he meant by this, John D. Rockefeller Jr. did so in a private letter to Mackenzie King:

> While, of course, we realized that we would have to deal with the unions so far as wages are concerned if the government so insisted, we have felt that there was no reason why the plan of representation should not be continued in both companies [CF&I and the Rockefeller's Consolidated Coal Company in Pennsylvania] and availed of to the fullest extent possible in connection with all other matters other than wages. In other words, it is our policy neither to fight the union where the government demands that it be recognized as the mouthpiece of the worker in the matter of wages nor to abandon the plan in all other matters where it can be availed of to the advantage of the workers and the company.[100]

At least in Colorado, the agreement did not work this way in practice. The contracts CF&I signed with the United Mine Workers of America included provisions dealing with much more than wages. Ironically, for the first time in decades, the mines operated without labor strife in the years following 1933. When production picked up later in the 1930s, CF&I coal miners still did not strike. The disputes that did occur were minor and short-lived. By 1937, even management was expressing satisfaction with the union agreement.[101] However, with no viable independent union alternative during most of the 1930s, the Rockefeller Plan continued to operate at the Minnequa Works until 1942.

Chapter Eight

DEPRESSION, FRUSTRATION, AND REAL COMPETITION

> [T]he success of the Joint Representation Plan within the Minnequa Plant has been largely due to the continual efforts on the part of the men and representatives to keep it morally clean. I have never considered our Plan a "paper plan." If these things which have come to mean so much to the employees, and which came to them through the Plan, are to be discontinued or modified, I think it will in some measure destroy the spirit and attitude of the men towards the Plan of Joint Representation within the Minnequa Plant of the C.F.&I. Co.
> —ANDREW DIAMOND TO JOHN D. ROCKEFELLER JR., AUGUST 23, 1933[1]

Throughout the 1920s, John D. Rockefeller Jr. took his family on various vacations in the American West and often stopped in southern Colorado. While visiting Pueblo in 1926, Rockefeller and his sons toured the Minnequa Works twice in the same day—once during the morning and once during the evening so the boys could see the plant in operation both during the day and at night, when the fires in the furnaces were more impressive.[2] Unlike his earlier trips in 1915 and 1918, few reporters followed Rockefeller on this visit. However, the *Industrial Bulletin* does mention that he met with a delegation of employee representatives from the steel mill, including the representatives' chair, Andy Diamond. At the next regular meeting of the workers who served under the employee representation plan (ERP), Diamond "stated that during the interview

DEPRESSION, FRUSTRATION, AND REAL COMPETITION

Mr. Rockefeller said he would be glad to have those present carry back to the men the information that he is more interested in the success of the Industrial Plan today than ever before, and to thank the representatives for the important part they play in making the Plan a success."[3] Almost seventy-five years later Rockefeller's son David recalled, "We spent a day in Pueblo touring the Colorado Fuel & Iron's large steel mills and meeting representatives of the company union. Father greeted a number of the men by name, and they seemed pleased to see him. I remember being a bit startled by the experience but impressed with my father's forthright manner and the easy way he dealt with the men and their families."[4] This warmth was a sign that the employee representatives at the steelworks still trusted Rockefeller to follow through on the promises he had made during the early days of the plan.

After the devastating strike of 1927–1928, however, Rockefeller's interest in employee representation at Colorado Fuel and Iron Company (CF&I) wavered. By September 1928 the company's stock had recovered from the low it experienced during that dispute and was higher than it had been in years. "While the company is doing very well," Rockefeller wrote Mackenzie King in 1928, "its future from an income producing point of view is still much in doubt. My associates have on various occasions urged me to sell my stock. . . . [W]ould you think it a serious mistake on my part, in so far as the Industrial Relations Plan, which you and I set up there, is concerned?"[5] Since Rockefeller chose not to sell, it seems reasonable to assume that King's answer was yes, even though no record exists of his response. This indicates at least some lasting support for the idea that employee representation could work at CF&I. His response to his son John D. Rockefeller III's 1929 Princeton senior thesis, "Industrial Relations Plans: A Study," also indicates Rockefeller's strong personal stake in the topic. "It has given me unspeakable pleasure to find the extent and breadth of your interest in the industrial relations problem," he wrote. "Your knowledge of the subject is far greater than I had imagined and many times more than mine at your age. It pleased me immensely too [sic] find that you so quickly saw the point of view which I have taken in these matters."[6] While this praise might be expected from a father to a son, the sentiments Rockefeller conveys nonetheless denote that his support for the plan was flagging because of the contents of that thesis.

John D. Rockefeller III criticized the Rockefeller Plan in terms similar to the way outside observers had attacked it just a few years earlier. For example, he wrote, "[I]t is hard to believe that the employees can become really enthusiastic over a plan which has no real power." Perhaps more extraordinary is John III's statement, "Mutuality of interest must exist before this or any other plan of industrial relations can be really successful," which attacks the foundation upon which King and Rockefeller built the plan.[7] This is tantamount to claiming that the plan had failed to achieve its most basic goal. Rockefeller granted his son an interview for the thesis, but he is not quoted in it. It is possible then that some of the criticism of the plan came from Rockefeller himself, since his son would have been unlikely to have risked estrangement from his father by criticizing his father's philosophy out of hand. Indeed, Rockefeller's silence on plan-related matters during this period speaks volumes. By 1933, preoccupied with other matters such as the building of Rockefeller Center and the growth of Colonial Williamsburg, he no longer concerned himself with matters pertaining to CF&I. According to an Associated Press story carried in *The Pueblo Chieftain* that year, Rockefeller had a "customary policy of non-interference with his companies."[8] Those in southern Colorado whose memories stretched back to 1915 knew this had not always been the case.

As Rockefeller lost interest in the plan, management began to administer it in such a way that it seemed more like the anti-union device its critics had always claimed it was. In 1928 CF&I hired Louis F. Quigg to serve as general superintendent at the Minnequa Works. Quigg came from Bethlehem Steel. By the late 1920s that firm's employee representatives were so intimidated that management initiated nearly every action by the representatives, and those it did not initiate invariably died in committee.[9] Despite Quigg's hostility toward representation, CF&I's employee representatives' interest in using the plan to their advantage grew over time. Elton Mayo noticed "a tendency to legalistic wrangling over the extent and literal meaning of a particular phrase or passage in the memorandum of agreement," reminiscent of the practices of trade unions in his native Australia.[10] For this reason, Quigg told the employee representatives at the Minnequa Works shortly after coming aboard, "The interpretation of [the] Plan should be in the hands of management, subject to criticism, review and discussion, but not necessarily subject to revision."[11] By 1930, employee representatives had so little power in the steel mill that they had to ask management to inform them about pending changes in the plant before they took effect.[12]

Management had always had the ability to control the plan this way; it chose not to do so, however, until the market for its products declined precipitously. Because of economic strains, company executives became increasingly willing to assert management's privileged position, especially as the employee representatives increasingly tested the limits of their power. The disputes grew more contentious as the Depression worsened. "I have no objection to saying yes [to a request] if the facts . . . justify that answer," Quigg explained to steelworks employee representatives at their annual joint meeting in 1931, "but I have never said no to so many who didn't believe me as I have out here."[13] Nevertheless, when a new independent union established a presence in Pueblo, the employee representatives and management again made common cause against the perceived interlopers.

A GREAT DEPRESSION FOR LABOR AND MANAGEMENT ALIKE

CF&I's business problems during the Depression dated back to the 1920s. Coal had difficulty competing with newly discovered petroleum-based products from Texas and Oklahoma. At the same time, a glut of coal on the market drove the price down to previously unheard-of lows. The market for steel rails (the company's most important product at the time) declined in 1921 as cars and trucks began to seriously compete with railroads for the first time in U.S. history. By the end of the decade, however, the firm's profits spiked in reaction to cuts in its coal operations and a stronger market for rails. CF&I began to modernize its Pueblo steel mill for the first time in twenty-five years, and the company experienced record earnings in the first quarter of 1930. Management even reinstated a dividend for stockholders, something it had not done in nine years.[14]

But when the Depression finally hit the company, it hit hard. By the end of 1930, demand in both sides of its business had dropped so precipitously that CF&I workers who remained on the payroll were drastically underemployed, working far fewer hours than they had in previous years. The details of this situation in the company's coal mines were discussed near the end of Chapter 7. The situation on the steel side of the business was almost as bad. At the 1931 annual meeting, a representative sent by President Arthur Roeder told the employee representatives, "We have some rail business, but it is just from one road, and considerably less than during the last two years. So there is nothing to be enthusiastic about. We would like you boys to bear in mind we have to trim our sails, and trim them closer than ever and watch expenditures."[15] By 1932, steel output at CF&I had dropped from 600,000 tons in both 1929 and 1930 to less than a third of that total.[16]

Fewer orders created severe economic hardship for steelworkers even if they retained their jobs. Management laid off many workers indefinitely, and even those who had work had fewer hours and hence smaller paychecks. Different departments of the Minnequa Works opened and closed on a week-to-week basis as orders for steel dictated. Even though Roeder recognized that CF&I steelworkers "want to work and don't want to stint on hours," the company still shared the available work among all the employees who remained on the payroll.[17] As the minutes of the annual joint meeting in 1935 explained, "[Superintendent] Quigg stated that so far as the company was concerned, their only purpose was to distribute the work." This policy inevitably irritated the highly paid skilled workers best represented by the ERP who sought overtime when work was available to make up for earlier times when the plant had been idle.[18]

In 1933, most steel companies turned a profit for the first time since the start of the Depression.[19] In contrast, CF&I chose not to pay interest due on the company's bonds in July 1933 so it would not have to sell its capital and could continue operating. As a result, it went into equity receivership.[20] The biggest holder of the bonds on which the company defaulted was John D. Rockefeller Jr. While it is impossible to pinpoint a single reason for the bankruptcy, the high labor costs stemming from the ERP could not have helped. The firm reorganized as the Colorado Fuel and Iron Corporation in 1936, although the Rockefeller interests remained the firm's largest stockholder. With significantly fewer operating mines than ever in its history, the new corporate entity was now primarily a steel company.[21]

Management's battles with its steelworkers during the 1930s revolved primarily around the benefits they had won under the auspices of the Rockefeller Plan. As early as 1925, John D. Rockefeller Jr.'s consultants from Curtis, Fosdick, and Belknap had complained of management's un-willingness to begin the "elimination of unnecessary jobs" at the steelworks and had argued that there should be "fewer men on the necessary jobs."[22] While labor constituted less than 20 percent of the total cost of CF&I steel, wages were still 31 percent higher than those at mills in the East (some jobs paid as much as 100 percent more than those of their eastern counterparts), thanks to the steelworkers' success at manipulating management through the ERP.[23] Continual lobbying by employee representatives kept wages comparatively high well into the 1930s. In July 1933, with the company's troubles at their highest point, President Roeder still announced that CF&I would guarantee its steelworkers "top wages in the industry."[24] While this strategy might have worked ten or twenty years earlier, the market in which CF&I operated had changed.

One of the biggest reasons for the drop in CF&I's steel business was the advent of eastern and foreign competition in its traditional market for the first time in the company's existence. Eastern and foreign steelworkers were paid substantially less than CF&I steelworkers; and thanks to a combination of desperation and improvements in transportation, companies from both areas had begun to poach on CF&I's market in the western United States.[25] Other companies might have shed workers during this challenging time, but the Rockefeller Plan, with its built-in protections that kept workers from being fired, made it much more difficult to remove high-paid senior workers from the payroll. Superintendent Quigg explained the problem bluntly in 1930: "[W]e have too many employees to begin with, and we have more men on our pay-rolls than it takes to operate this plant economically by trying to take care of them."[26] Since management had been hiring and firing less-skilled workers on practically a biyearly basis for over a decade, Quigg had to have been referring specifically to skilled white native-born workers with long tenures and high salaries—the ones most likely to take advantage of the plan and the kind of workers who tended to serve as employee representatives.

THE STEEL WORKS YMCA

In the wake of the bankruptcy, CF&I management shut down the Steel Works YMCA in August 1933 as a cost-saving measure. It did so under a court order issued as a result of its receivership. Built in 1920 with a $500,000 gift from John D. Rockefeller Jr., the building had been a significant drain on the company's finances ever since.[27] (The "Y" buildings in the coal towns kept operating because they were much less expensive to run.) Ironically, the money CF&I had used to run the building went to the company's bondholders instead—most notably John D. Rockefeller Jr., who had given the money to build the YMCA in the first place. Understandably, the talk in Pueblo when the "Y" closed was that Rockefeller would step in and save the building he had allowed to be built. At a special meeting on August 10, called to discuss the fate of the company under receivership, Roeder agreed to have the court delay the closing of the "Y" until September 1, 1933, so a delegation of employees led by Andrew Diamond could make a direct appeal to keep the building open. "The inspiration for such approach," explained *The Pueblo Chieftain*, "is the frequently demonstrated personal interest upon the part of Rockefeller in the employees of the C.F.&I."[28] As a stockholder, Rockefeller could not make up the deficit to keep the building open because of the receivership, but he could have done so as a private citizen. However, his son Nelson feared that had his father decided to fund the "Y" as a gift, he would have been asked to make many similar gifts.[29] While there was truth in that position, it was not what the employee representatives wanted to hear from management.

Employee representatives took the news of the "Y's" closing very hard. Earlier that year, Employee Representative Carl Thompson, chair of that year's membership drive, had declared, "The CF&I Steel Works Y has fully met an important demand of thousands of steelworkers and their families in providing educational, social and athletic diversion at a time when all other sources have failed."[30] Later in the campaign, Diamond told the *CF&I Blast*, "Because the employees themselves through their representatives asked for the Steel Works Y, it is their duty to support it to the fullest extent."[31] Yet as was the case with the Rockefeller Plan itself, some steelworkers supported the YMCA more than others. The mill departments with the greatest number of memberships after the 1933 campaign were the Wire Mill, the office staff, and the Open Hearth—two departments with many highly skilled workers and one made up of professionals. Departments in which everybody joined the YMCA that year included the telegraph office, the blueprint room, the general superintendent's

DEPRESSION, FRUSTRATION, AND REAL COMPETITION

8.1. *The Steel Works YMCA, Pueblo, CO. Courtesy, Bessemer Historical Society.*

office, and industrial relations.[32] These highly skilled workers and management staff had the money to afford memberships for themselves and their families; workers desperate for hours did not.

Membership in the "Y," like participation in the Rockefeller Plan itself, reflected economic and therefore racial divisions within CF&I's workforce. Management understood this dynamic well. "If the Steel Works YMCA was endowed so that it was free to all steelworkers and their families," President Roeder wrote to a Rockefeller aide in 1936 in a discussion of how much John D. Rockefeller Jr. should donate to keep the "Y" operating, "we would immediately face the race and color problem and also the hoodlum element would get out of control."[33] With countless Mexican and Mexican American workers laid off because of the Depression, the race and color problem was worse than ever before. These workers had not frequented the "Y" when steadily employed; management did not want them there while they were unemployed.

In contrast, the skilled white native-born workers' fondness for the YMCA is the reason the company's decision to close the building was so important for the operation of the Rockefeller Plan. The closing adversely impacted the constituency in the plant that had always made the greatest use of the ERP, people like Andy Diamond. Diamond and his fellow representatives sent Rockefeller a telegram immediately after the closing announcement appealing to him to keep the YMCA open. They argued specifically that since he had built the "Y" in response to requests made by representatives under the plan, the "Y" was in fact part of the plan. Rockefeller offered no response. Diamond then wrote Rockefeller a personal letter. "It seems peculiar that you could inject yourself into the management affairs of the company several years ago," he told him, "and now at a time when more serious problems confront the employees you take the position that nothing can be done by you to assist the employees at this time. I feel it was a grave mistake to tell a committee of employees that you would meet with them at any time."[34] He was probably referring to their 1926 meeting. Diamond, at least, had taken Rockefeller's promises at the time seriously, and now he was bitter about their failure to materialize.[35]

Diamond had told many other steelworkers that his personal relationship with the industrialist would help employee representatives when they needed it. Now that Rockefeller had not come through in a difficult time, Diamond was, as Roeder put it, "in a rather awkward position."[36] Even the *Chieftain* had reported that "Diamond is a friend of the multi-millionaire."[37] Rockefeller's failure even to respond to Diamond's letter likely only added insult to injury. Instead, Rockefeller's aide Robert Gumbel sent Diamond a telegram stating that "Rockefeller and the stockholders generally have already made large contributions through the foregoing of dividends during the greater part of the last 20 years."[38]

Even before the Great Depression began, John D. Rockefeller Jr. was almost solely responsible for keeping the democratic promise of the Rockefeller Plan alive. When Rockefeller's interest in employee representation lapsed in the 1930s, so did the success of the plan in maintaining harmony within CF&I. "The Rockefeller Plan will not work as well with Rockefeller back in New York as it would if he remained here," predicted John Lawson back in the plan's early years.[39] He was mostly right. As long as Rockefeller took an active interest in the ERP, his liberal management style defined its operation. When he stopped caring, the considerably more traditional men who ran the Minnequa Works undermined his original vision, and the employees responded angrily. Part of that anger revolved around the end of welfare capitalism at CF&I.

Historians have devoted considerable space to the benefits of welfare capitalism during the 1910s and 1920s, but only a few have considered that anger could result when welfare efforts stopped. Lizabeth Cohen has explored this possibility in her study of Chicago during the New Deal: "[T]he majority of industrial workers felt abandoned by the welfare capitalists.... Traditional elites at the workplace and in the community, on whom workers had depended during the previous decade, seemed to be letting them down."[40] Yet employee anger at CF&I went well beyond the removal of these programs.

EMPLOYEE REPRESENTATIVES' OTHER FRUSTRATIONS

Tired of having to press so hard for their once-traditional rights under the plan, employee representatives at the Minnequa Works grew increasingly belligerent as the difficult circumstances brought on by the Depression made even the most highly skilled workers fear for their jobs. Superintendent Quigg called these workers "pugnacious."[41] This attitude is best exemplified by a single case. In May 1934 management discharged first helper James Anderson for having allowed excessive heat to run through the wall of an open hearth furnace, thereby ruining the expensive piece of equipment. Anderson was a U.S. citizen who had been with the company since 1920. During that period he had worked his way up to third helper and then quickly to first helper.[42] According to Anderson and the employee representatives who took his side, he was not even on the job when the incident occurred. As Andrew Diamond explained at the August 1, 1934, joint ERP meeting, "What was done with the man from the out-start [sic] of the case was that he was incompetent. That is what management proved, or attempted to prove. From the evidence that was brot [sic] into us we didn't see it that way." During the months of discussion involving this case, the employee representatives often asked out loud why, if Anderson was really incompetent or unqualified for the position, he had obtained it in the first place. The significance of the case derived from its role as a symbol for management's prerogatives under the plan. "I championed the Plan as I did the first time I signed up for it," Diamond told Roeder at the same meeting. "But if management attempts to interpret to their advantage certain paragraphs in there ... they are not [being] fair."[43] Since management had fired Anderson under these circumstances, other skilled workers—maybe even employee representatives—could have met the same fate.

Anderson used his rights under the plan to appeal his dismissal. Diamond shepherded Anderson's case through the appeals process all the way up to President Roeder. Roeder upheld the dismissal. From there, Diamond tried to bring the case before an arbiter, but management refused to allow it. Instead, management brought the case before the Colorado Industrial Commission, which decided that the charges against Anderson had not been proven by testimony. Indeed, the commission suggested that Superintendent Quigg's charge that all open hearth furnace tenders at the plant were "careless . . . in handling open hearth bottoms" was "a serious indictment of open hearth workers generally."[44] Unfortunately for Anderson, a recent Colorado Supreme Court decision had declared that commission decisions were binding only if both parties gave their consent, which management refused to do.[45]

The case ended up before the judge overseeing CF&I's bankruptcy. Although the judge declared that management had not proven its case against Anderson, he refused to overturn the decision.[46] Representative James Irwin, later vice president of the ERP under Andrew Diamond, summed up the result of the Anderson case shortly after the judge's decision. "I want to say that Jim Anderson was abused," he told his fellow representatives, "and is still being abused these last five or six months. You can talk about how good the Plan is, but it hasn't done a thing for Jim Anderson."[47] At least one employee representative, Richard Dolan, resigned his position because he felt he was being discriminated against as a result of his actions in Anderson's defense.[48]

The Anderson case ultimately brought home the limits of employee freedom under the plan even to Andrew Diamond. This exchange between Diamond and Superintendent Quigg occurred at the April 18, 1935, meeting of the Joint Committee on Industrial Cooperation, Conciliation, and Wages, almost a year after Anderson's discharge:

> Quigg: I think it is stated in those minutes that the aggrieved party could take it [the Anderson dispute] to the Commission, but neither party had to abide by the Commission's findings. . . .
>
> Diamond: If your interpretation is right about the Plan and the State Industrial Commission, then it is the biggest joke that was ever put on a group of working men.

> Quigg: You forget the earlier statement in your Plan and which you all agreed to, that the Plan is a written agreement between the men and management. You forget the statement and pay no attention where it specifically states that management reserves the right to select and to direct its working forces and nothing in the Plan can abridge that right. . . . No business can run with an outsider running it.[49]

Quigg used the word "outsider" in reference to his own employees, not to some walking delegate based in New York City. It must have been a terrible blow to a man who professed to believe in John D. Rockefeller Jr.'s philosophy, as Diamond did, to learn that labor and management were not in fact "partners."

While representatives such as Diamond would ultimately choose the Rockefeller Plan over an independent union, their continued willingness to press management suggests that they did not do so out of loyalty to their employers. Instead, they saw that they could stand up for their own interests more effectively in the structure they had shaped over the previous twenty-plus years than they could in a new organization. This is exemplified in a 1934 debate over wage increases coinciding with a brief upturn in the company's steel orders. While management wanted to give workers an average wage increase of 10 percent (some departments more, some less), Diamond and the employee representatives wanted the raise to be 10 percent across the board. "Under joint representation, the interpretation means to treat one and all the same, and that's what should be done," declared Diamond at an April 25 meeting. "Now when we first started out with our Plan all our raises and all our reductions were given either a straight 10% or a straight 15% and there really wasn't any adjustments [sic]."[50] Diamond left unstated the fact that since most employee representatives were the highest-paid men in the plant, their wages would no doubt have been on the low end rather than the high end of management's variable range. Despite Quigg's protestations that workers would get more time if they held labor costs down, he gave in to the representatives' demand.[51] As long as the plan stayed in effect, highly paid workers had the potential to gain further victories. However, starting in 1936, the power of skilled employees at the mill faced a serious threat from a new kind of steel union.

NEW UNION, OLD PLAN

"Fundamental principles of the C.F.&I. joint representation plan will be the basis for similar plans in other major industries in the country and provided for in the text of the National [Industrial] Recovery Act signed this week by President Franklin D. Roosevelt," boasted the *CF&I Blast* in June 1933.[52] The act, whose Section 7(a) guaranteed workers the right to bargain collectively through representatives of their own choosing for the first time in U.S. history, had inspired the drive that led to the United Mine Workers of America's (UMWA) contract with CF&I coal miners, but it did little to change the situation for CF&I steelworkers. The Amalgamated Association of Iron, Steel and Tin Plate Workers did little in response to the passage of the act; in fact, the only reason it did anything at all was that it was prodded by the American Federation of Labor to take advantage of a prime opportunity to organize more locals.[53] The passage of the National Labor Relations Act (NLRA) in 1935 replaced the labor provisions that had been thrown out with the provisions of the unconstitutional National Industrial Recovery Act. It also created the National Labor Relations Board (NLRB) to enforce its provisions. This in turn led to a new push for union organizing across the country, spearheaded by the Committee (later Congress) of Industrial Organizations (CIO), which emphasized industrial union organization. That meant it included every worker in an industry rather than the skill-based craft organizing the American Federation of Labor had always favored.

Many of the new worker organizations that sprang up in response to the passage of the National Industrial Recovery Act (NIRA) and the National Labor Relations Act were company-initiated and company-dominated. As one U.S. Department of Labor survey concluded in 1937, "[A] large number of company unions—more than half . . . performed none of those functions which are usually embraced under the term 'collective bargaining.' Some of these were merely agencies for discussion. Others had become essentially paper organizations after their primary discussion was performed when a trade union was beaten."[54] Such organizations have continued to give employee representation a bad name to the present day. However, they were not the only kind of ERP created during the 1930s.

DEPRESSION, FRUSTRATION, AND REAL COMPETITION

Although there were more than a few bad examples, a surprisingly large number of these new organizations offered the same kinds of rights to employees as those provided under the Rockefeller Plan. Economist David Saposs explained:

> A brand new type of company union is coming into existence simultaneously with the readaptation of old ones. This one approaches the characteristics and collective bargaining practices of trade-unions. Unlike the readapted plans, this new type of "company union," of which a few existed in pre-NRA [National Recovery Administration, set up by the NIRA] days, is organized as a separate entity. It has a clear-cut constitution. It also has a full-fledged written trade agreement covering wages, hours, and other working conditions clearly resembling that entered into between employers and trade-unions.[55]

Economist John Pencavel has found a closer correlation between "company unions" and higher wages in the 1930s than he saw in shops without such organizations.[56] That is another indicator of the similarity between some "company unions" and independent trade unions. These new ERPs did not constitute widespread recognition that John D. Rockefeller Jr. was right about the benefits of industrial democracy. Instead, they came about because of Section 8(a)(2) of the NLRA, which made it illegal for an employer to "dominate or interfere with the formation or administration of any labor organization or contribute financial or other support to it."[57] These employers were willing to pay employees more and to foster labor organizations that resembled independent trade unions in their shops to hold off the newfound threat of independent unions.

The creation of the Steel Workers Organizing Committee (SWOC) of the CIO also played an important role in this new wave of organizing. In fact, the AFL's failure to organize steel was one of the most important reasons the UMWA and other unions left to form the CIO in 1935. The SWOC was formed on June 3, 1936, when the old Amalgamated Association accepted a deal with the CIO that gave SWOC complete control over a new organizing campaign to be conducted along industrial lines. Only with these developments did a movement to establish an independent union at the Minnequa Works truly take hold. The SWOC's top-down structure had an important effect on the campaign in Colorado, as it depended more on the hiring of experienced organizers than it did on support from the rank and file.[58] This explains why rather than organizing being limited to skilled workers (as had been the case with the Amalgamated Association in the past), steelworkers of all skill levels joined SWOC because the vast majority of SWOC organizers came from the UMWA.[59] In line with its reputation as a racially enlightened organization, the CIO actively solicited Mexican and Mexican American steelworkers to join the union, even drawing up fliers in Spanish.[60] Skilled white workers like Andrew Diamond opposed SWOC because when it fought for the interests of these workers it upset his power base in the ERP and, by consequence, the power relationships in the entire plant.

Yet just how much power any group had at the Minnequa Works remained unclear for years. During the late 1930s, it was common for workplaces across the country to have two, three, or even four trade unions vying for the employees' loyalty. The Minnequa Works had three: the Steel Workers Organizing Committee, the old Rockefeller Plan, and another independent union called the Western States Steel Products Union, which never had more than a few hundred members. Many workers held cards in more than one organization. This was not the kind of situation the politicians behind the NLRA had imagined. The purpose of Section 8(a)(2) was to force employees to choose between independent unions and nothing at all. No matter how munificent an ERP might have been toward employees, such organizations became illegal. The Rockefeller Plan as operated to that point in time ran afoul of Section 8(a)(2) under this interpretation. Therefore, the company initiated two sets of changes in the plan to allow it to pass muster as an independent union.[61] In 1937, an employee representative suggested further revisions to the plan as a result of the Supreme Court decision in *NLRB vs. Jones and Laughlin Steel Corporation*.[62] Contrary to most people's expectations, that decision upheld the constitutionality of the National Labor Relations Act. Therefore, a new version of the plan had to be passed to guarantee that it could meet NLRB scrutiny.[63]

Despite these changes, the Rockefeller Plan remained essentially the same as the one Mackenzie King had written almost twenty-five years earlier, even after a second set of changes. The committee structure survived intact, as did key clauses such as the one requiring nondiscrimination in union membership. The new plan did clarify that if the representatives of labor and management could not resolve their differences in the appeals process all the way through the president of the company, either party could force the other to submit to independent arbitration. This policy was designed to avoid a repeat of the Anderson case. In the end, the parties to the agreement initiated these changes to purge the plan of management's direct role in operations, although that role remained intact in the accompanying Memorandum of Agreement. The National Labor Relations Board eventually rejected the notion that any distinction existed between the plan itself and the Memorandum of Agreement and noted that "only minor changes were made in the Plan" at this juncture. The NLRB also noted that the 1937 changes were not submitted to employee representatives and that rank-and-file employees were not "informed that they were free to change the form of representation which had been initiated by the Company."[64] Management used this approach because it did not believe the ERP had enough popular support to allow it to compete with an independent union the way John D. Rockefeller Jr. had originally imagined.

In addition to the structure of the new plan, another indicator of its fundamental resemblance to the old Rockefeller Plan was the similarity in the types of people who served as employee representatives during the plans' respective periods of operation. Table 8.1 lists the employee representatives who signed the 1938 version of the plan. At least six had served as employee representatives during the previous decade and, based on the representatives from 1938 whose personnel cards could be found, had also started to work for the company during that decade. Many had been with CF&I for as much as twenty years. In short, this list reflects the continued, even greater dominance of the ERP by a white, native-born, skilled elite within the steel plant over time. Since the Steel Workers Organizing Committee threatened these workers' power base, it directed its organizing campaign toward everyone else.

THE ROCKEFELLER PLAN VS. THE STEEL WORKERS ORGANIZING COMMITTEE

Although SWOC began organizing in Pueblo in 1937, the first SWOC lodge did not form there until August 23, 1938.[65] Its campaign did not gain momentum until the first professional organizer arrived in March

Table 8.1. Employee representatives at Minnequa Works, 1938

Name	Nationality/Race	First Year w/ CF&I	Position in 1938
Tony Amato	Italian	1916	labor—14" mill
Clarence Banks	African American	1916	head switch man
Herman J. Becker	American	1912	brickmason
Preston G. Crocker	American	1929	crane man
A. A. Cunning	*	(first elected) 1918	*
R. D. Davidson	American	1918	third helper open hearth
Edgar R. DeCeer	*	*	*
Andrew J. Diamond	American	1914	first assistant roller
Michael G. Filler	*	(first elected) 1925	*
Joe Guadagno	American	1937	trucker
E. C. Haggerty	[citizen]	1917	inspector
Gus J. Hoffman	*	*	*
John F. Hoover	*	*	*
James Irwin	American	1924	millwright
W. F. Kaufman	American	1919	engineer
Charles W. Kelsay	American	1936	crane man
Anton Kochevar	Austrian	1911	water tender (boiler)
Jim Lacy	*	*	*
Fred Lenza	*	*	*
John J. Masciota	[citizen]	1913	spell levers
Chas. McGill	*	*	*
John Pagano	*	*	*
Joe W. Perko	Bohemian	1913	heater helper
John Piserchio	[illegible]	1917	rope maker
Hiram W. Prior	*	(first elected) 1925	roll turner
M. J. Rosevach	[illegible]	1911	header
Chas. F. Schreiber	*	*	*
B. D. Selman	*	*	*
Burl Soots	*	*	*
R. W. Speakman	*	*	*
Edward Thomas	American	1913	foreman
W. H. Thompson	American	1926	bundler
Frank Van Dyke	American	1917	first helper open hearth
Joseph K. Whalen	*	*	*
Ted G. White	*	*	*
Alvin L. Woods	*	*	*

* Information is not available.
Signatories from Memorandum of Agreement, Employee Representatives Organization, April 1, 1938.

1939. Even then, the organizer quarreled with local leadership for another year, so no real progress was made. After Fred Hefferly, son of District 15 head and state CIO president Frank Hefferly, arrived in 1940, membership grew quickly.[66] It is possible to track the progress of the union in the mill by counting dates on the card SWOC submitted to the NLRB in support of its first representation election: 192 bore the date 1937, 490 were dated 1938, 219 were from 1939, and 553 had the date 1940.[67] By 1940, SWOC had new offices and a stewardship system under way in the plant and claimed to represent a majority of the workers.[68] This example illustrates the two primary tracks SWOC took both in Pueblo and around the country: it organized workers on the ground and simultaneously used the machinery provided by the National Labor Relations Act. One strategy reinforced the other.

The testimony collected by the NLRB examiner who came to Pueblo to investigate the SWOC complaint included scores of incidents in which CF&I management and its representatives had violated the rights of SWOC members and sympathizers. For example, as SWOC employee organizer Paul Ducic testified:

> Well, I passed . . . [Mike] Filler [employee representative] the particular machine he was operating on this day, and Filler came to me and said, "Paul, I understand that you are organizing for the C.I.O." And I could not deny the fact.
>
> I told him that I had [been doing so]; and he asked me how many we had signed up during the time that I had went about with the local C.I.O. organizer, and I attempted to . . . tell him the best I could the circumstances at the time of what we had done, and Filler then . . . mentioned the name[s] of other men in the mill that were organizing for the C.I.O., and particularly a man in the coke plant, and that these men were liable to eventually get into trouble because the company would not stand for any union—would not stand for any C.I.O. to be located in the C.F. and I. plant.

Ducic did not testify that Filler had threatened to punish him for organizing a union but reported that Filler had stated that CF&I would fire Ducic "for possibly coming intoxicated on the job or making bad weaves of wire on these poultry netting machines, or any number of minor mistakes that might possibly occur."[69] It is therefore no surprise that another witness testified before the NLRB that "there is a lot of fellows [who] fear that if they would join the C.I.O. that they might be discharged. . . . [E]very time we talked things over . . . these representatives admitted . . . that I ought to know that the company wouldn't stand for the C.I.O."[70] As the membership numbers cited earlier demonstrate, this intimidation worked. SWOC only began to gain a foothold in the plant after the NLRB ruled against the ERP in 1940.

To grow the union, SWOC directly attacked the credibility of the employee representatives serving under the plan, especially Andrew Diamond. Clearly, Diamond had been in power for so long that he had made many enemies. The CIO organizers must have drawn on those enemies when they drew up a series of leaflets that portrayed Diamond in extremely unflattering terms. The flier with the longest text attacking Diamond preserved in the Frank and Fred K. Hefferly Papers in the University Archives of the University of Colorado at Boulder cited history to make SWOC's case:

> The record of Mr. Jack is too long for us to forget. He has served his master well. It dates back to the strike of 1919, when Jack organized a back to work movement of scabs to break the strike.
>
> He then became a "big shot" for the Rockefeller plan. Messrs. Diamond and Irwin never moved without the approval of the company. The steel workers have paid the bill in increased profits gotten by the company. The company always named the representatives and picked the attorneys.
>
> The workers have always been wise to the role of Mr. Diamond, who serves as an "Uncle Tom" to his masters of industrial slavery. For years the company was able to use fear and force to intimidate those who would organize for their own protection.[71]

Other leaflets attacking Diamond portray him as a carnival barker, as a crying baby, and as the Statue of Liberty being toppled by the district court decision upholding the NLRB's dissolution order for the ERP. Faced with such personal criticism, it is easy to imagine why Diamond and other employee representatives fought SWOC tooth and nail despite their own frustrations with management.

The SWOC fliers also took pains to argue that the Rockefeller Plan had failed in its obligations to steelworkers. "The company union is powerless to adjust minor grievances as has been proven many times," read one leaflet. "Take for example the Anderson case which is familiar to nearly every employee of the company."[72] Another leaflet directly attacked the ERP's claims of repeated victories for the rank and file:

> We wonder how many workers have received two weeks vacation with pay. Exactly none. The eight hour day was in force in many industries long before Andy ever came to the fair city of Pueblo and it was not a company union that put it into effect here, but a direct action of the government at that time. You who have worked in the mill know what happens when you are called for and [sic] emergency shift. Have you ever collected time and a half for it?[73]

Yet while making such arguments in an effort to build its organization, SWOC continued to use the new machinery created by the National Labor Relations Act to ensure that the majority it gathered could express itself freely in a government-run representation election.

THE ERO AND ERO, INC.

Even before it began organizing in earnest, the Steel Workers Organizing Committee asked the new National Labor Relations Board to invalidate the Rockefeller Plan so organizing would be easier. SWOC's first complaint, filed before the NLRB on March 30, 1938, charged that CF&I "dominated and interfered" with the ERP and "contributed financial and other support to it" in violation of the NLRA. In April, the board ordered an investigation that occurred in July. An NLRB trial examiner spent a week in Pueblo taking testimony. In October, the examiner found in favor of the union. CF&I immediately appealed to the federal NLRB in Washington, D.C.[74]

Among the witnesses at the national NLRB hearing on the CF&I case were ERP chair Andrew Diamond and vice chair James Irwin. During his testimony Diamond offered an impassioned defense of the plan and of his actions under its auspices. "It is certain we are not dominated [by management]," he told the NLRB. "We would not even belong to a company that could dominate us." NLRB chair Warren Madden did not take such protestations by a "company union" member seriously. "The men elected under the plan had a very good thing out of it," he remarked during the hearing, "something which was financially very profitable to them."[75] Madden was referring to the fact that the representatives were paid for attending meetings, but he might also have pointed to the power Diamond and Irwin had achieved under the auspices of the plan. Their mere presence at a hearing in Washington, D.C., attests to the fact that they were different from other steelworkers, who did not have the same opportunity to travel and still maintain their salaries.[76]

On March 29, 1940, the federal NLRB issued its decision against the Rockefeller Plan. The basis of the decision was the fact that the ERP continued to be a "creature" of management even after the post-NLRA changes, which therefore invalidated it under Section 8(a)(2) of the NLRA. "Not only was the Plan the same organization after the 1938 revisions as it formerly had been," the NLRB wrote, "but no one regarded the revisions as establishing a new organization. The 'by-laws' of the revised Plan describe it as a revision of the Plan adopted in 1916. [Andrew] Diamond, speaking from the point of view of the employee representatives, testified that there had only been one Plan with revisions." The NLRB also noted that "[t]he employee representatives, chosen prior to the revisions, continued as employee representatives under the Plan as revised. The Industrial Bulletin containing the Proceedings of the meetings under the Plan, and still published by the respondent [CF&I] for distribution to all employees, was numbered consecutively without any break in sequence after June 1937."[77] This decision applied to both the steel plant in Pueblo and the iron ore mine in Sunrise, Wyoming, which faced a parallel organizing campaign against the ERP there by the International Union of Mine, Mill and Smelter Workers.[78]

The National Labor Relations Board's decision did not, however, force management to recognize the Steel Workers Organizing Committee as the bargaining unit in the plant. Even as management appealed the NLRB decision to a federal court, the union asked for and received permission to conduct an NLRB-run election at the Minnequa Works over three days in March 1941. The former employee representatives under the Rockefeller Plan did not give up easily, however. They changed the name of their group to the ERO (Employee Representative Organization) and encouraged steelworkers to vote against both the SWOC and the weak Western States Steel Products Union. "A Vote for Neither is a Vote for ERO," was the ERP slogan.[79] When the votes were tallied, SWOC received 1,738 votes, the other union received 319 votes, and the category "Neither" received 2,670 votes.[80]

DEPRESSION, FRUSTRATION, AND REAL COMPETITION

After the vote, Andrew Diamond issued a statement declaring victory, as well as his intention to proceed as if nothing had changed within the plant.[81] In fact, it had. In the end, the people whose interpretation of the election result mattered most were the staff and members of the National Labor Relations Board. In the official notice announcing the election, the NLRB had written, "You cannot vote for any labor organization not named on the ballot by writing in its name *or in any other manner*" (emphasis added).[82] With this warning, the ERO should have realized that any "victory" it achieved in the election would be short-lived. Certainly SWOC recognized that fact, or else it would have campaigned in an entirely different manner. In his report on the election, CIO organizer Frank Bonacci noted that SWOC never challenged the ERO's notion that "a vote for NEITHER would be a vote for the E.R.O., Company Union."[83] SWOC's decision to undo that mistake explains why the union refused to accept the election results. "The SWOC-CIO received 1,783 votes which means it is the predominant union in [the] plant," read SWOC's statement after the election. "We intend to gain recognition and representation for our membership."[84]

After the election, CF&I asked the Tenth Circuit of the Federal District Court to vacate the NLRB order disestablishing the ERP. SWOC argued otherwise, and the court took the union's position. It reaffirmed the national NLRB order in a June 23, 1941, decision. The court ruled that the influence of the old form of representation on the election had tainted the result to such an extent that it had to be abolished to ensure freedom of choice by employees.[85] In response to the order affirming the disestablishment of their union, ERO leaders filed incorporation papers, turning the ERO into ERO, Inc., a nominally independent organization. In fact, the articles of incorporation for the ERO were drawn up by CF&I's lawyer, A. T. Stewart.[86] In August 1941 this "new" union presented management with 4,000 signed membership cards and asked that it be recognized as the bargaining agent in the plant. Management agreed.[87]

The new agreement between CF&I and ERO, Inc., bore a much greater resemblance to the kinds of union contracts workers were just beginning to sign in the war industries than it did to the Rockefeller Plan. It was certainly an improvement over the last version of the ERP. The contract included double pay for employees called to the plant for emergencies; paid holidays on Christmas, the Fourth of July, and Labor

Day; unpaid leaves of absence; and perhaps most important for the steelworkers who ran the ERO, respect for seniority. "Promotions shall be made on the basis of seniority," read clause VII(4). "Men with seniority where practicable shall be given the preference for preferred days off," read VII(5). Perhaps the surest sign that the steelworkers had won the battle they had started in the 1920s was the layoff procedure written into the contract. The priorities governing layoffs listed seniority at the top (to protect longtime workers from layoffs). Employee qualifications came in second.[88] However, the agreement would not be in force long enough to test this provision.

In response to the agreement with ERO, SWOC filed another complaint against management with the NLRB. The board agreed that the contract violated Section 8(a)(2) of the NLRA, and on June 8, 1942, the Tenth District Court of Appeals upheld that order. It ruled that although management had tried to cut its ties with the old ERO, it had still given the new organization a slight advantage over the independent union. It ordered the company to withdraw its recognition of ERO, Inc., and call a new election.[89] In response to that ruling, the SWOC began an intense organizing campaign at the plant. As plant manager Louis Quigg later recalled:

> Immediately after the newspaper account broke, every boy in the plant was interested. It became a subject of conversation throughout the plant, and shortly thereafter the SWOC flooded the plant with what I call dodgers, calling attention to the employes that the appellate court had thrown out the old organization; they were no longer the collective bargaining agency; and they started an intensive drive through advertising with their own dodgers and activities in the plant.[90]

During the same month in which the court dissolved ERO, Inc., SWOC adopted its first constitution. That document changed the name of the union to the United Steel Workers of America (USWA).[91] When the new NLRB-ordered representation election occurred in late July 1942, the USWA received 2,723 votes in favor to 1,886 against.[92] The union and the company signed their first agreement on October 1, 1942. As Thomas Hogle has explained, "[P]roponents of the Rockefeller Plan never again mounted an effort to replace the union. The company's steel workers came to rely on the union as had the coal miners in 1933."[93]

8.2. *The National Labor Relations Board election that led to the eventual certification of the United Steel Workers of America as the exclusive bargaining unit of CF&I steelworkers, 1942.*

At the opening of the new seamless tube mill in 1953, CF&I distributed a small round knife as a souvenir of the event. One side of the case depicts labor and management holding a shared olive branch. This was the embodiment of labor-management cooperation the Rockefeller Plan could not provide.[94] While much of the steel industry disappeared with the advent of global competition in the 1960s, Local 2102 survives to the present day. In fact, it won the longest strike in USWA history in 2004.[95] The record of John D. Rockefeller Jr.'s ERP pales in comparison.

Rockefeller himself tacitly acknowledged the superiority of independent unions over his plan through his actions. Despite the collapse of the plan in the Colorado coal fields, Rockefeller still stated to the man in charge of Rockefeller Center in 1938: "I believe that employers and employees are partners, not enemies. That their interests are essentially common interests, not antagonistic. That the highest well-being of both can best be obtained by cooperation, not by warfare."[96] Nevertheless, as he gradually retired from all business activities during the 1930s, Rockefeller allowed his sons to turn their backs on his paternalistic labor policies when they took control of management of the family's empire.[97] More important for the present history, John D. Rockefeller Jr. sold his family's interest in the Colorado Fuel and Iron Company to a consortium headed by Wall Street financier Charles Allen Jr. in 1945. With the company stock having rebounded because of government contracts during World War II, it was the perfect time to sell. The parties announced the deal two days before Christmas in 1944. The president of the company at the time, W. A. Maxwell, told the Associated Press that he had received no warning of the deal. When the AP contacted Rockefeller's office, his representatives responded, "[T]here is no statement forthcoming."[98] Rockefeller's associates said nothing either. Clearly, the family intended to attract as little attention to the sale as possible, lest someone remember episodes in the company's history that John D. Rockefeller Jr. would have preferred that everyone forget.

CONCLUSION

> Men versed in the tenets of freedom become restive when not allowed to be free.
> —SENATOR ROBERT WAGNER OF NEW YORK, MARCH 11, 1934[1]

Increasingly uninterested in his earlier efforts at labor relations reform as the decades passed, John D. Rockefeller Jr. never acknowledged that his experiment in industrial relations at the Colorado Fuel and Iron Company (CF&I) had gone wrong. The Rockefeller Plan did not bring labor peace to the company, a failure that hurt its financial position at a time when CF&I could ill afford any setbacks. A 1949 letter from a disgruntled former CF&I stockholder, Howard Briggs, explains the inevitable result of labor trouble in two highly competitive, increasingly depressed industries:

> I am writing this letter to call to your attention something which was written some 30 years ago, and which impressed me at that time as being exceptionally fine. The booklet I refer to was entitled—"The Three Legged Stool." One leg represented the public, the second labor and the third ownership. This would have been a wonderful theory had it been carried out! However, as you know, the leg representing ownership was either cut off, or fell off the stool. In my 40 years of ownership (or over) I cannot recall having ever received anything from the common stock.[2]

CONCLUSION

While he misremembered some of it, Briggs was clearly referring to Rockefeller's speech inaugurating the ERP in Pueblo back in 1915.

If Rockefeller ever read Briggs's complaint, he should have sympathized with his position. After all, nobody suffered more financially from the losses and subsequent failure of CF&I than he did. However, the memory of Rockefeller's early career as a labor reformer would also have been painful for him then. As Daniel Okrent has written, "In Junior's view, consistency was an essential component of honor; hidden from his view but embedded within his psyche was an unimaginative man's need to cling to consistency as if to a life raft."[3] This explains why Rockefeller stuck with employee representation at CF&I much longer than proved financially prudent. He wanted to put people before profits, yet when reality finally caught up with Rockefeller's utopian outlook, he simply moved on to other projects he could better control. Rockefeller's position on labor reform was consistent to the end, but when both CF&I and the Rockefeller Plan that operated there turned sour, he stopped showing any interest in labor reform.

Despite the failure of Rockefeller's experiment, the story of the Rockefeller Plan still offers an important lesson in industrial relations: employee representation plans (ERPs) can help bring about independent unions rather than hinder them. "[E]ven in unsympathetic hands," Gary Dean Best has argued in reference to the World War I era, "the employee representation movement opened the door to education in industrial democracy and the wider concept of collective bargaining as many of its advocates intended that it should."[4] Writing about the 1930s, David Brody has concluded, "The evidence suggests that ERPs did foster local leadership and, insofar as they failed to produce results, did educate workers and strengthen the case for collective bargaining by outside unions."[5] CF&I faced this particular problem because workers in both sides of the company's business wanted more than management was willing to provide. However, had the Rockefeller Plan given employees the freedom and agency independent unions provided, it would have ceased to have been employer-dominated. The resultant frustration made all firms with "company unions" highly susceptible to outside organizing. It also made it easy for the United Mine Workers of America (UMWA) to organize at CF&I when it had government protection during its 1933 revival.

CONCLUSION

I have tried to avoid the term "company union" in this work largely because the phrase suggests only one kind of non-union employee representation: organizations in which management dominates labor. Like the term "anti-Federalists" used during the debate over the U.S. Constitution in the late 1780s, "company union" is a name invented by the detractors of that idea.[6] The use of that term indicates acceptance of the critique without examining it; in fact, calling an ERP that is not employer-dominated a "company union" is a contradiction in terms. Not all ERPs were "company unions" in this classic sense, and U.S. workers who have dealt with them recognize that fact. CF&I workers viewed the Rockefeller Plan the same way John Pencavel does, seeing the ERP as part of a spectrum "with completely independent unions at one end of that range and employer-dominated organizations at the other."[7] Even without offering freedom and independence, a flawed union proved better for CF&I workers than no union at all because it became the vehicle by which management heard labor's voice and bent (at least in part) to its employees' will. The workers who received benefits from the plan saw half a loaf as better than nothing.

One reason for this attitude may have been that the workers recognized that an employee representation plan's position along the continuum between independent unions and employer-dominated organizations was not stable over time. Circumstances change, often for the worse, thereby offering an opportunity for independent unions to build on the precedent for bargaining that ERPs may have already set. It took time for CF&I workers to replace the Rockefeller Plan with an independent union, but this delay can largely be explained by the fact that employee representatives made the arrangement work for them, at least to a limited degree. However, as the market for CF&I's steel changed during the Great Depression and management became more intent on controlling its workforce and limiting the scope of the ERP, many employees found the ERP increasingly inadequate.[8]

In today's extremely low-density union environment, an employee representation plan that resembles the Rockefeller Plan should be welcome since even a "company union" would be an improvement over nothing, which is what most American workers now have. Yet neither the union movement nor the legal system has kept up with these changes. Indeed, the law with respect to "company unions" in the United States has stayed the same over the past fifty years: "[p]ristine," to use David Brody's phrase, "without the usual encrustation of amendment and case law of a sixty-year old provision."⁹ However, to justify suspending National Labor Relations Act (NLRA) 8(a)(2) and related provisions of the Wagner Act in amber, the justification for opposition to "company unions" has changed drastically. John D. Rockefeller Jr.'s solution to the labor question was that workers should be able to choose the best bargaining unit for their own interests. This idea did not sit well with the labor movement then, but with unions so weak today, it should be more welcome. Indeed, the history of the Rockefeller Plan can actually teach us a great deal about the relatively recent debate over Section 8(a)(2) of the NLRA, an issue that has moved to the forefront of labor law reform in recent years.

"COMPANY UNIONS" AND THE LAW

The example of the Rockefeller Plan helped inspire a flurry of new employee representation plans during the 1920s. Many were explicitly intended as anti-union devices. In opposition to this tendency, the 1926 federal Railway Labor Act designated that railroad workers had a right to organize "without interference, influence, or coercion" over their "designation of representatives." When workers on the Texas and New Orleans Railroad (T&NO) chose the Brotherhood of Railway and Steamship Clerks to represent them, the railroad created and recognized an ERP. In 1930 the U.S. Supreme Court decided the case of the *Texas and New Orleans Railroad Company v. Brotherhood of Railway and Steamship Clerks* with concerns about an active labor movement in mind, since skilled railway workers had the strongest unions in America. Nevertheless, Chief Justice Charles Evans Hughes, writing the 8–0 majority opinion, explained the reasoning behind the provision by which the Court upheld a lower-court decree invalidating the ERP at T&NO:

> Freedom of choice in the selection of the representatives is the essential foundation of the statutory scheme. All the proceedings looking to amicable adjustments and to agreements for arbitration of disputes, the entire policy of the act, must depend for success on the uncoerced action of each party through its own representatives to the end that agreements be satisfactorily reached and the peace essential to the

CONCLUSION

uninterrupted service of the instrumentalities of interstate commerce may be maintained.[10]

So the Court's main reason for upholding this provision of the Railway Labor Act was to prevent strikes. It strongly suggests that ERPs could lead to strikes, since unhappy union supporters would not accept them as alternatives to independent trade unions. Most noteworthy, the Depression and the subsequent union revival that accompanied it had barely started when this decision was written. Hughes recognized that hostility toward ERPs could make the labor movement active even at a time when it was still largely dormant, as it had been in the 1920s.

Section 7(a) of the 1933 National Industrial Recovery Act (NIRA) is perhaps the best-known clause in that famous yet extinct law because it marked the first time the U.S. government had recognized the right of workers to "organize and bargain collectively through representatives of their own choosing." However, this landmark provision also contained language specifically designed to combat "company unions." "[N]o employee and no one seeking employment shall be required as a condition of employment to join any company union," it read. As was the case in the T&NO Railroad decision, the justification for the NIRA as a whole was economic—namely, to prevent "widespread unemployment and disorganization of industry, which burdens interstate and foreign commerce."[11] If businesses could have coerced their workers to join ERPs to circumvent Section 7(a), the ensuing strikes would have aggravated rather than helped the economic emergency, hence the explicit language in the act.

The 1935 National Labor Relations Act (NLRA) contains similar language in Section 8(a)(2).[12] (It also contains other provisions that support this section.)[13] Differences of opinion over the nature of non-union ERPs constituted much of the debate over the NLRA in the U.S. Congress. As briefly noted in Chapter 3, Senator Robert Wagner of New York was the primary author and architect of this legislation. What is seldom noted, however, is the similarity between Wagner's goals with respect to worker organization and those of John D. Rockefeller Jr. As Wagner explained during a hearing prior to the passage of the act, his bill did not prevent ERPs and independent unions from competing "in an open field." If this makes no sense in light of Wagner's inclusion of Section 8(a)(2) in the

NLRA, recall the principle behind the T&NO Railroad decision: government wanted workers to be happy with their unions so there would be no strikes. Any "company union" that workers chose freely would likely keep them happy on the job. Where Wagner and Rockefeller differed was that Wagner thought that a truly free choice between ERPs and independent unions could only occur when the government protected workers to offset the inherent power held by management in any employment relationship.[14] As Wagner explained to service station employee John Collins during hearings for the NLRA, "All I am trying to do, and I think, you believe me when I say that, is to make the worker a free man to join any organization that he wishes to join and, at the same time, have genuine collective bargaining."[15] Only independent unions could provide genuine collective bargaining under the law, otherwise Section 8(a)(2) would have been unnecessary.

By 1939, the U.S. Supreme Court had carried Wagner's philosophy to its logical extreme. In *National Labor Relations Board v. Newport News Shipbuilding & Dry Dock Co.*, the Court upheld an NLRB order even though it agreed that the company's employees had freely chosen the ERP as their bargaining representative. "In applying the statutory test of independence," explained the majority opinion, "it is immaterial that the plan had, in fact, not engendered, or indeed had obviated, serious labor disputes in the past, or that any company interference in the administration of the plan had been incidental, rather than fundamental and with good motives."[16] In other words, the Court was recognizing that congressional will superseded employee choice with respect to ERPs. It left the workers with the option of an independent union or nothing. The *Newport News* decision was an important precedent for the federal court that upheld the invalidation of the Rockefeller Plan in 1942. In that decision, the ERP's long history became a liability rather than an asset, since it tainted the ability of the United Steel Workers to compete against it on a fair playing field.

As the NLRB limited workers' choices to independent unions or no unions at all, the number of ERPs declined quickly. Nevertheless, after the passage of the NLRA, the NLRB worked to almost completely eradicate what it saw as "company unions" from the American workplace. Between 1935 and 1940, the NLRB disestablished 700 such organizations, 502 in fiscal year 1940–1941 alone.[17] By gaining winner-take-all representation elections under the auspices of the NLRA, the U.S. labor movement grew at a remarkable rate through World War II. Since independent unions won over 85 percent of NLRB representation cases between 1935 and 1945, these organizations came to favor NLRB-administered representation elections as their favorite mode of organizing.[18] But organized labor's power in U.S. society declined precipitously shortly thereafter. By the 1980s, American workers often voted against representation of any kind (even though those votes were no longer fair, as they once had been). Workers then had no organization left to turn to—not even a "company union."

As employer resistance to trade unions has increased in recent decades, many non-union workers have been left with non-union shops as their only option and no real hope for organization. Many employers fiercely resist unionization, and anti-union administrations have appointed NLRB members who have little interest in protecting workers' rights under the NLRA.[19] After George W. Bush was reelected in 2004, labor lawyer Thomas Geoghegan stopped just short of begging for an outright repeal of the law. "The act is so screwed up," he wrote, "management could hardly have it any better. Worried about the National Labor Relations Board? Not really. No union serious about organizing uses it anymore."[20] Despite this farcical situation with respect to organizing rights, Section 8(a)(2) of the NLRA still prevents independent unions from using "company unions" as organizing vehicles.

Indeed, NLRB decisions regarding organizing rights are not the only decisions that have become controversial. One fairly recent precedent strongly suggests that workers lack the intelligence to distinguish between half-measures in collective bargaining and truly independent unions. In 1989, Electromation, Inc., an electrical parts manufacturer, established its first joint labor-management action committees in response to employee concerns over wages. The Teamsters union soon began an organizing campaign against the firm and filed an NLRB complaint against the action committees. The NLRB upheld a judge's ruling that the committees were "labor organizations" under the language of the NLRA and ordered them disbanded because the employer "in their formation and administration unlawfully supported them."[21] The premise of the *Electromation* decision is that workers who benefited from the labor-management committees would have no need for independent labor unions. As the majority opinion stated, "The purpose of the Action Committees was . . . not to enable management and employees to cooperate and improve 'quality' and 'efficiency,' but to create in employees the impression that their disagreements with management had been resolved *bilaterally*" (emphasis in original). In other words, the NLRB proceeded as if management had instituted the committees to fool workers into thinking their problems had been addressed even if those problems remained unsolved. A concurring opinion in *Electromation* is even more direct, suggesting that the action committees "gave employees the illusion of a bargaining representative without the reality of one."[22] The assumption here is clear: workers cannot recognize true collective bargaining for themselves. In the 1993 *Du Pont* decision, which the NLRB used to clarify its ruling in *Electromation*, the board stated, "The threshold question for the determination of whether an employer has violated section 8(a)(2) is whether the entity involved is a labor organization."[23] Therefore, all forms of nonunion employee representation are now automatically deemed illegal, regardless of whether they are employer-dominated.[24] The two decisions are still in force today.

The result of the long line of reasoning in these two cases is that employers cannot legally create labor organizations under the NLRA; only workers can. There is no middle ground. Employer disgust over the *Electromation* and *Du Pont* decisions helped inspire the first serious proposed reform to Section 8(a)(2) (and its supporting provisions in the NLRA) since the original passage of the act in 1935. In 1996, Congress passed the TEAM (Teamwork for Employees and Managers) Act. If President Bill Clinton had not vetoed it, the TEAM Act would have weakened NLRA rules regarding employer-dominated labor organizations, making it possible for employers "to establish, assist, maintain or participate in any organization or entity of any kind" that addresses "matters of mutual interest, including, but not limited to, issues of quality, productivity, efficiency, and safety and health."[25] Union supporters expressed strong fears that changes such as this would have significantly undercut workers' right to organize. Writing about the TEAM Act in 2000, Jonathan P. Hiatt and Laurence E. Gold of the AFL-CIO's Legal Department explained their concern that the bill would lead to "an explosion of employer-sponsored organizations that deal with terms and conditions of employment but stop short of meaningful employee representation."[26] This critique assumes that such arrangements would be stable. Like the NLRB in *Electromation*, Hiatt, Gold, and other supporters of Section 8(a)(2) do not seem to entertain the possibility that American workers can tell an independent union from a company-dominated organization and might strike in response to any less than independent organization when the opportunity became available.

The history of the Rockefeller Plan demonstrates that at least some workers could tell a good union from a bad one. CF&I workers who participated in the plan turned an organization with glaring flaws into something that could be used to forward their interests—increased wages, shorter work hours, and other benefits. Through this process, the Rockefeller Plan paved the way for future support for independent trade unionism, offering employee representatives solid reasons to seek the intangible qualities related to the kind of freedom of action only outside organizations could offer. They were helped in achieving this objective by changes in the wider world, most notably a Depression that led to a Democratic presidential administration that gave workers the legal right to organize for the first time in U.S. history. Government protection for the right to organize proved crucial in helping to ensure that the industrial democracy John D. Rockefeller Jr. had created actually reflected the will of the company's employees.

"SOMETHING ABSOLUTELY DEMOCRATIC"

In recent years, some observers who are anything but anti-labor have been willing to entertain the notion of amending (or even repealing) the National Labor Relations Act. Some of this reformist zeal has centered on Section 8(a)(2). For example, in reaction to *Electromation*, the U.S. Commission on the Future of Worker-Management Relations (better known as the Dunlop Commission) recommended in 1994 that "nonunion employee participation programs [not be found] unlawful simply because they involve discussion of 'terms and conditions' of work or compensation as long as such discussion is incidental to the broad purposes of these programs."[27] The commission made that suggestion as part of a proposed deal to improve workers' ability to organize.[28] Similarly, Bruce Kaufman has suggested that "the ban on 'company unions' was not a wise policy decision at that time [the 1930s], nor does it serve the public interest, or even the interests of organized labor at the current time."[29] However, he is careful to advocate changes in the NLRA's language dealing with employer-dominated labor organizations only if such changes are accompanied by assurances that the government would protect the right of workers to organize independent unions if they so desire. These changes, he argues, would protect the interests of both employers and employees.[30] Raymond Hogler and Guillermo Grenier have made a similar suggestion: repeal 8(a)(2), but in exchange further modify the NLRA to allow unionization through card check in lieu of an election.[31] This bears some resemblance to the system suggested under the Employee Free Choice Act, a bill labor unions are currently working hard to have enacted.

Even in the absence of state support to protect workers' organizing power, there should be no question that employees were better off belonging to a paternalistic ERP such as the Rockefeller Plan than they were with no union at all. Paternalistic, solicitous employee representation plans at least offer workers something. Why should the law say that workers must accept all or nothing when it comes to unions? John Logan, in recounting the history of the TEAM Act, has explained that U.S. labor leaders' major concern about the bill was not "that employees would be 'duped' into seeing employer-dominated representation as something other than it was; rather, their concern was that employees who wanted unions would often settle for a lesser form of workplace voice because of its easy accessibility in a system that exacted a high price for independent representation."[32] But why should workers not be able to decide independently what is best for them?

Many workers cannot expect the option of an independent union anytime soon. The idea of organized labor balancing the power of organized business in America today is almost laughable. Workers have enough trouble trying to survive; making the perfect the enemy of the good by rejecting the possible benefits to workers of a non-union ERP is not a smart idea today as it was in 1935. Many new kinds of workers today have neither the prospect nor a history of union representation. As Sanford Jacoby has explained:

> Some, including myself, are deeply concerned that none of these proposals [from the Dunlop Commission] addresses the needs of employees who are outside career-type employment relationships.

CONCLUSION

> The Commission's proposals focus on enterprise representation, yet recent years have seen partial disintegration of the stable employment relationships that are a prerequisite to enterprise representation. Large nonunion companies like IBM and Eastman Kodak no longer guarantee quasi-lifetime jobs. A rapidly increasing share of the workforce is comprised of contingent employees (part-time, temporary, and contract employees) whose ties to the employer—and whose employers' ties to them—are weak.[33]

Similarly, an employee representation plan at a fiercely anti-union firm like Walmart might be the only option for representation those workers ever have. Mainstream independent trade unions have shown no interest in organizing these workers. Why should they? It is difficult enough to organize permanent workers in today's labor relations climate. Besides, if these workers elect to join some type of employee representation plan, they are not necessarily lost to the labor movement forever. Since nonunion employee representation plans have often served as springboards for independent unions, it is easy to see that workers' decision to accept a less than independent union need not be permanent.

Perhaps the best way to judge the legitimacy of labor organizations, independent and dependent alike, actually came from John D. Rockefeller Jr. The goal he established for his plan was industrial democracy. "Our thought has been to devise something absolutely democratic," he told reporters during his first visit to Colorado in 1915, "something that will take in all workmen whether they belong to the union or not." In the same interview, he called his plan "broader and more democratic than unionism."[34] The term "democracy" has been poorly served by its overuse in the corporatist literature in recent years.[35] However, it has not always been this way. As Nelson Lichtenstein has explained, the idea of "industrial democracy" was once a powerful concept for both workers and managers. During the 1940s and 1950s, however, it "practically vanished from the vocabulary of political life and union discourse."[36] By allowing this to happen, the labor movement essentially left the high moral ground of democracy and workers' rights to its enemies.

As a result of this retreat, reformers now have trouble distinguishing legitimate labor reform from management-inspired conservative propaganda. Ian Sakinofsky has offered an excellent criterion with which to judge whether true democracy is present in a particular workplace:

CONCLUSION

> True worker voice, whether exercised directly within organizations or indirectly using such third party representatives as trade unions, provides those workers about whom decisions are being made or for whom the decisions have consequence, with the ability to modify those decisions. It is thus the potential extent, of employee input into decision making, that is quite likely the truest measure of real worker voice—whether this input be made within a union system or within a nonunion system.[37]

Participants in the Rockefeller Plan did have their voices heard. They frequently affected and even changed management decisions. However, many workers felt they affected too few decisions for their liking and therefore continually expressed dissatisfaction with the system. If a plan as solicitous of, and beneficial to, its employees as the Rockefeller Plan did not keep them happy, it is difficult to see how any non-union ERP could have done so. If such organizations returned to the American workplace, it seems unlikely that the labor movement would be in worse shape than it is now. The worst-case scenario would be that a few non-union workers would move a little further down the continuum from non-union despotism toward an independent union, but they would still not attain their goal of such a union.

Rockefeller's goal of creating "something absolutely democratic" was ahead of its time. Indeed, the notion of true industrial democracy in the United States might *still* be ahead of its time, as most workers still do not have it. As Hogler and Grenier have explained, "Rockefeller's philosophy of mutual interests has been absorbed into a textured ideology of corporate munificence, democracy and individual fulfillment through organizational design."[38] While better than nothing, this is not the basis of a just and equitable society. A book about history is not the best place to suggest ways to change the present, but my work on the Rockefeller Plan has led me to believe that the preconditions for such changes would not necessarily be entirely economic. Cultural changes that cease to define workplace democracy as coming only from independent unions would make intermediate unions not associated with the labor movement at large more palatable. Furthermore, political reform, particularly new legislation protecting the right to organize, would help protect members of employee representation plans from the kind of management domination that destroyed the reputations of these organizations during the 1930s.

CONCLUSION

What the Rockefeller Plan lacked during most of its history was an active state, present (even if only in the background) to protect workers as they exercised their right to choose the union they saw fit. Whenever labor can be easily replaced, management always has the upper hand in the workplace. This was certainly the case in the coal and steel industries during the years in which the Rockefeller Plan operated. Only government could balance management's inherent advantage, and no impartial government arrived on the scene until the 1930s. Even then, President Franklin Roosevelt's goal was not necessarily to promote independent unions. He famously told reporters in 1934 that he did not care if workers joined unions, the Ahkoond of Swat, or the Royal Geographic Society.[39] It was the workers' prerogative, not the president's, as to what kind of organizations they wanted, and Congress passed Section 7(a) of the National Industrial Recovery Act (the law that provoked this remark) to make sure workers could exercise that right however they saw fit. If Roosevelt trusted workers to choose their representatives wisely, why shouldn't we? After all, the people most likely to understand the interests of American workers are the workers themselves.

The notion that democratic non-union employee representation plans can be good vehicles for organizing independent unions should not be controversial. In addition to the historical evidence, considerable contemporary examples from across the world of unions "capturing" management-dominated organizations exist. In Canada, for example, one-third of the members of the Communication, Energy and Paperworkers Union used to belong to "company unions" or independent associations.[40] In Australia, Parliament reconfigured the Workplace Relations Act in 1996 to permit non-union collective agreements under Section s170LK. Rae Cooper and Chris Briggs have found instances in which these agreements "enabled unions to 'jump start' organizing campaigns in non-union workplaces."[41] A similar provision could have an even greater impact on organized labor in the much less unionized United States. Wolfgang Streeck, writing about Germany, has pointed out that "more than ever, industrial unions use the works council as the institutional framework and the major source of support for their activities at the workplace and in the enterprise."[42] This kind of international comparison should be helpful for an American labor movement willing to entertain new ways to achieve its goal of obtaining new members. Like corporate campaigns and community-

CONCLUSION

based unions, the return of employee representation plans could be another effective method to revive organized labor in the American workplace, assuming the democratically endowed right to organize receives adequate protection.

APPENDIX 1: THE COLORADO INDUSTRIAL PLAN (ALSO KNOWN AS THE ROCKEFELLER PLAN) AND THE MEMORANDUM OF AGREEMENT

THE INDUSTRIAL CONSTITUTION

PLAN OF REPRESENTATION OF EMPLOYES IN THE COAL AND IRON MINES OF THE COLORADO FUEL AND IRON COMPANY OF COLORADO AND WYOMING

I

REPRESENTATION OF EMPLOYES

1. Annual meetings for election of employes' representatives.

Employes at each of the mining camps shall annually elect from among their number representatives to act on their behalf with respect to matters pertaining to their employment, working and living conditions, the adjustment of differences, and such other matters of mutual concern and interest as relations within the industry may determine.

---

* Source: John D. Rockefeller Jr., "The Colorado Industrial Plan" (1916), 63–94. This pamphlet was privately printed by John D. Rockefeller Jr. and distributed to important people around the country with a card that read "COMPLIMENTS OF JOHN D. ROCKEFELLER, JR." For references to other versions of the plan, see Chapter 3, note 12.

2. Time, place, and method of calling annual meetings, and persons entitled to be present and participate in the election of representatives.

The annual meetings of employes for the election of their representatives shall be held simultaneously at the several mining camps on the second Saturday in January.

The meetings shall be called by direction of the president of the company. Notices of the meetings, indicating their time and place, as well as the number of representatives to be elected, shall be publicly posted at each camp a week in advance, and shall state that employes being wage-earners in the employ of the company at the time of the meeting and for at least three months immediately preceding, but not salaried employes, shall be entitled to be present and vote. Special meetings shall be similarly called when removal, resignations, or other circumstance occasions a vacancy in representation.

3. Method of conducting meetings, and reporting election of representatives.

Each meeting for the election of employes' representatives shall choose its own chairman and secretary. At the appointed hour, the meeting shall be called to order by one of the employes' representatives, or, in the absence of a representative, any employe present, and shall proceed to the election of a chairman and secretary. The chairman shall conduct, and the secretary record, the proceedings. They shall certify in writing to the president of the company the names of the persons elected as the employes' representatives for the ensuing year.

4. Basis and term of representation.

Representation of employes in each camp shall be on the basis of one representative to every one hundred and fifty wage-earners, but each camp, whatever its number of employes, shall be entitled to at least two representatives. Where the number of employes in any one camp exceeds one hundred and fifty, or any multiple thereof, by seventy-five or more, an additional representative shall be elected. The persons elected shall act as the employes' representatives from the time of their election until the next annual meeting, unless in the interval other representatives may, as above provided, have been elected to take their places.

5. *Nomination and election of representatives.*

To facilitate the nomination and election of employes' representatives, and to insure freedom of choice, both nomination and election shall be by secret ballot, under conditions calculated to insure an impartial count. The company shall provide ballot boxes and blank ballots, differing in form, for purposes of nomination and election. Upon entering the meeting, each employe entitled to be present shall be given a nomination ballot on which he shall write the names of the persons whom he desires to nominate as representatives, and deposit the nomination ballot in the ballot box. Each employe may nominate representatives to the number to which the camp is entitled, and of which public notice has been given. Employes unable to write may ask any of their fellow employes to write for them on their ballots the names of the persons whom they desire to nominate; but in the event of any nomination paper containing more names than the number of representatives to which the camp is entitled, the paper shall not be counted. The persons—to the number of twice as many representatives as the camp is entitled to—receiving the highest number of nomination votes shall be regarded as the duly nominated candidates for employes' representatives, and shall be voted upon as hereinafter provided. (For example: If a camp is entitled to two representatives, the four persons receiving the largest number of nomination votes shall be regarded as the duly nominated candidates. If the camp is entitled to three representatives, then the six persons receiving the largest number, etc.)

6. *Counting of nomination and election ballots.*

The chairman shall appoint three tellers, who shall take charge of the ballot box containing the nomination votes, and, with the aid of the secretary, they shall make out the list of the duly nominated candidates, which shall be announced by the chairman. The meeting shall then proceed to elect representatives by secret ballot, from among the number of candidates announced, the same tellers having charge of the balloting. If dissatisfied with the count, either as respects the nomination or election, any twenty-five employes present may demand a recount, and for the purposes of the recount the chairman shall select as tellers three from the number of those demanding a recount, and himself assist in the counting, and these four shall act, in making the recount, in place of the secretary and the tellers previously chosen. There shall be no appeal from this recount, except to the president of the company, and such appeal may be taken as hereinafter provided, at the request of any twenty-five employes present and entitled to vote.

7. Appeal in regard to nomination or election.

The chairman of the meeting shall preserve for a period of one week both the nomination and election ballots. Should an appeal be made to the president within seven days in regard to the validity of the nomination or election, upon a request in writing signed by twenty-five employes present at the meeting, the chairman shall deliver the ballots to the president of the company for recount. Should no such request be received within that time, the chairman shall destroy the ballots. If after considering the appeal the president is of the opinion that the nomination or election has not been fairly conducted, he shall order a new election at a time and place to be designated by him.

8. General proceedings at meetings.

At annual meetings for the election of representatives, employes may consider and make recommendations concerning any matters pertaining to their employment, working or living conditions, or arising out of existing industrial relations, including such as they may desire to have their representatives discuss with the president and officers of the company at the Annual Joint Conference of the company's officers and employes, also any matters referred to them by the president, other officers of the company, the Advisory Board or Social Joint Committee appointed at the preceding annual joint conferences of officials and employes of the company. A record of the proceedings shall be made by the secretary of the meeting and certified to by the chairman, and copies delivered to each of the representatives, to be retained by them for purposes of future reference.

II

DISTRICT CONFERENCES, JOINT COMMITTEES AND JOINT MEETINGS

1. District divisions.

To facilitate the purposes herein set forth, the camps of the company shall be divided into five or more districts, as follows: the Trinidad District, comprising all mines and coke oven plants in Las Animas County; the Walsenburg District, comprising all mines in Huerfano County; the Canon District, comprising all mines in Fremont County; the Western District, comprising all mines and coke oven plants located on the Western Slope; the Sunrise District, comprising the iron mines located in Wyoming.

2. *Time, place and purpose of district conferences.*

District conferences shall be held in each of the several districts above mentioned at the call of the president, at places to be designated by him, not later than two weeks following the annual election of representatives, and at intervals of not more than four months thereafter, as the operating officers of the company, or a majority of the representatives of the employes in each of the several districts, may find desirable. The purpose of these district conferences shall be to discuss freely matters of mutual interest and concern to the company and its employes, embracing a consideration of suggestions to promote increased efficiency and production, to improve working and living conditions, to enforce discipline, avoid friction, and to further friendly and cordial relations between the company's officers and employes.

3. *Representation at district conferences.*

At the district conferences the company shall be represented by its president or his representative and such other officials as the president may designate. The employes shall be represented by their elected representatives. The company's representatives shall not exceed in number the representatives of the employes. The company shall provide at its own expense appropriate places of meeting for the conferences.

4. *Proceedings of district conferences.*

The district conferences shall be presided over by the president of the company, or such executive officer as he may designate. Each conference shall select a secretary who shall record its proceedings. The record of proceedings shall be certified to by the presiding officer.

5. *Joint committees on industrial relations.*

The first district conferences held in each year shall select the following joint committees on industrial relations for each district, which joint committees shall be regarded as permanent committees to be intrusted with such duties as are herein set forth, or as may be assigned by the conferences. These joint committees shall be available for consultation at any time throughout the year with the Advisory Board on Social and Industrial Betterment, the president, the president's executive assistant, or any officer of the operating department of the company.

(a) Joint Committee on Industrial Cooperation and Conciliation: to be composed of six members.

(b) Joint Committee on Safety and Accidents: to be composed of six members.

(c) Joint Committee on Sanitation, Health and Housing: to be composed of six members.

(d) Joint Committee on Education and Recreation: to be composed of six members.

6. *Selection and composition of joint committees.*

In selecting the members of the several joint committees on industrial relations, the employes' representatives shall, as respects each committee, designate three members and the president of the company or his representative, three members.

7. *Duties of Joint Committee on Industrial Cooperation and Conciliation.*

The Joint Committee on Industrial Cooperation and Conciliation may, of their own initiative, bring up for discussion at the joint conferences, or have referred to them for consideration and report to the president or other proper officer of the company at any time throughout the year, any matter pertaining to the prevention and settlement of industrial disputes, terms and conditions of employment, maintenance of order and discipline in the several camps, company stores, etc.

8. *Duties of Joint Committee on Safety and Accidents.*

The Joint Committee on Safety and Accidents may, of their own initiative, bring up for discussion at the joint conferences, or have referred to them for consideration and report to the president or other proper officer of the company at any time throughout the year, any matter pertaining to the inspection of mines, the prevention of accidents, the safeguarding of machinery and dangerous working places, the use of explosives, fire protection, first aid, etc.

9. Duties of Joint Committee on Sanitation, Health and Housing.

The Joint Committee on Sanitation, Health and Housing may, of their own initiative, bring up for discussion at the joint conferences, or have referred to them for consideration and report to the president or other proper officer of the company at any time throughout the year, any matter pertaining to health, hospitals, physicians, nurses, occupational disease, tuberculosis, sanitation, water supply, sewage system, garbage disposal, street cleaning, wash and locker rooms, housing, homes, rents, gardens, fencing, etc.

10. Duties of Joint Committee on Education and Recreation.

The Joint Committee on Education and Recreation may, of their own initiative, bring up for discussion at the joint conferences, or have referred to them for consideration and report to the president or other proper officer of the company, at any time throughout the year, any matter pertaining to social centers, club houses, halls, playgrounds, entertainments, moving pictures, athletics, competitions, field days, holidays, schools, libraries, classes for those who speak only foreign languages, technical education, manual training, health lectures, classes in first aid, religious exercises, churches and Sunday schools, Y.M.C.A. organizations, etc.

11. Annual and special joint meetings.

In addition to the district conferences in each of the several districts, there shall be held in the month of December an annual joint meeting, at a time and place to be designated by the president of the company, to be attended by the president and such officers of the company as he may select and by all the employes' representatives of the several districts. At this meeting reports covering the work of the year shall be made by the several joint committees and matters of common interest requiring collective action considered. A special joint meeting of any two or more districts may be called at any time upon the written request to the president of a majority of the representatives in such districts or upon the president's own initiative, for the consideration of such matters of common interest as cannot be dealt with satisfactorily at district conferences.

Notice of such special joint meetings shall be given at least two weeks in advance.

III

THE PREVENTION AND ADJUSTMENT OF INDUSTRIAL DISPUTES

1. Observance of laws, rules and regulations.

There shall be on the part of the company and its employes, a strict observance of the federal and State laws respecting mining and labor and of the company's rules and regulations supplementing the same.

2. Posting of wages and rules.

The scale of wages and the rules in regard to working conditions shall be posted in a conspicuous place at or near every mine.

3. No discrimination on account of membership or non-membership in labor or other organisations.

There shall be no discrimination by the company or by any of its employes on account of membership or non-membership in any society, fraternity or union.

4. The right to hire and discharge, and the management of the properties.

The right to hire and discharge, the management of the properties, and the direction of the working forces, shall be vested exclusively in the company, and, except as expressly restricted, this right shall not be abridged by anything contained herein.

5. Employes' right to caution or suspension before discharge.

There shall be posted at each property a list of offenses for commission of which by an employe dismissal may result without notice. For other offenses, employes shall not be discharged without first having been notified that a repetition of the offense will be cause for dismissal. A copy of this notification shall, at the time of its being given to an employe, be sent also to the president's industrial representative and retained by him for purposes of future reference. Nothing herein shall abridge the right of the company to relieve employes from duty because of lack of work.

APPENDIX I

Where relief from duty through lack of work becomes necessary, men with families shall, all things being equal, be given preference.

6. Employes' right to hold meetings.

Employes shall have the right to hold meetings at appropriate places on company property or elsewhere as they may desire outside of working hours or on idle days.

7. Employes' right to purchase where they please.

Employes shall not be obliged to trade at the company stores, but shall be at perfect liberty to purchase goods wherever they may choose to do so.

8. Employes' right to employ checkweighmen.

As provided by statute, miners have the right to employ checkweighmen, and the company shall grant the said checkweighmen every facility to enable them to render a correct account of all coal weighed.

9. Employes' right of appeal to president of company against unfair conditions or treatment.

Subject to the provisions hereinafter mentioned, every employe shall have the right of ultimate appeal to the president of the company concerning any condition or treatment to which he may be subjected and which he may deem unfair.

10. Duty of president's industrial representative.

It shall be the duty of the president's industrial representative to respond promptly to any request from employes' representatives for his presence at any of the camps and to visit all of them as often as possible, but not less frequently than once every three months, to confer with the employes or their representatives and the superintendents respecting working and living conditions, the observance of federal and State laws, the carrying out of company regulations, and to report the result of such conferences to the president.

11. Complaints and grievances to be taken up first with foremen and superintendents.

Before presenting any grievance to the president, the president's industrial representative, or other of the higher officers of the company, employes shall first seek to have differences or the conditions complained about adjusted by conference, in person or through their representatives, with the mine superintendent.

12. Investigation of grievances by president's industrial representative.

Employes believing themselves to be subjected to unfair conditions or treatment and having failed to secure satisfactory adjustment of the same through the mine superintendent may present their grievances to the president's industrial representative, either in person or through their regularly elected representatives, and it shall be the duty of the president's industrial representative to look into the same immediately and seek to adjust the grievance.

13. The right of appeal to the superior officers of the company against unfair treatment, conditions, suspensions or dismissals.

Should the president's industrial representative fail to satisfactorily conciliate any difference, with respect to any grievance, suspension or dismissal, the aggrieved employe, either himself or through his representative—and in either case in person or by letter—may appeal for the consideration and adjustment of his grievance to the division superintendent, assistant manager or manager, general manager or the president of the company, in consecutive order. To entitle an employe to the consideration of his appeal by any of the higher officers herein mentioned, the right to appeal must be exercised within a period of two weeks after the same has been referred to the president's industrial representative without satisfactory redress.

14. Reference of differences in certain cases to Joint Committees on Industrial Cooperation and Conciliation.

Where the president's industrial representative or one of the higher officials of the company fails to adjust a difference satisfactorily, upon request to the president by the employes' representatives or upon the initiative of the president himself, the difference shall be referred to the Joint Committee on Industrial Cooperation and Conciliation of the district and the decision of the majority of such joint committee shall be binding upon all parties.

15. Representation on joint committees to be equal when considering adjustment of differences.

Whenever a Joint Committee on Industrial Cooperation and Conciliation is called upon to act with reference to any difference, except by the consent of all present the joint committee shall not proceed with any important part of its duties unless both sides are equally represented. Where agreeable, equal representation may be effected by the withdrawal of one or more members from the side of the joint committee having the majority.

16. Umpire to act with joint committees in certain cases.

Should the Joint Committee on Industrial Cooperation and Conciliation to which a difference may have been referred fail to reach a majority decision in respect thereto, if a majority of its members so agree, the joint committee may select as umpire a third person who shall sit in conference with the committee and whose decision shall be binding upon all parties.

17. Arbitration or investigation in certain cases.

In the event of the Joint Committee on Industrial Cooperation and Conciliation failing satisfactorily to adjust a difference by a majority decision or by agreement on the selection of an umpire, as aforementioned, within ten days of a report to the president of the failure of the joint committee to adjust the difference, if the parties so agree, the matter shall be referred to arbitration, otherwise it shall be made the subject of investigation by the State of Colorado Industrial Commission, in accordance with the provisions of the statute regulating the powers of the commission in this particular. Where a difference is referred to arbitration, one person shall be selected as arbitrator if the parties can agree upon his selection. Otherwise there shall be a board of three arbitrators, one to be selected by the employes' representatives on the Joint Committee of Industrial Cooperation and Conciliation in the district in which the dispute arises, one by the company's representatives on this committee, and a third by the two arbitrators thus selected.

By consent of the members of the Joint Committee on Industrial Cooperation and Conciliation to which a difference has been referred, the Industrial Commission of the State of Colorado may be asked to appoint all of the arbitrators or itself arbitrate the difference. The decision of the sole arbitrator or of the majority of the Board of Arbitration or of the members of the State of Colorado Industrial Commission when acting as arbitrators, as the case may be, shall be final and shall be binding upon the parties.

18. Protection of employes' representatives against discrimination.

To protect against the possibility of unjust treatment because of any action taken or to be taken by them on behalf of one or more of the company's employes, any employes' representative believing himself to be discriminated against for such a cause shall have the same right of appeal to the officers of the company or to the Joint Committee on Industrial Cooperation and Conciliation in his district as is accorded every other employe of the company. Having exercised this right in the consecutive order indicated without obtaining satisfaction, for thirty days thereafter he shall have the further right of appeal to the Industrial Commission of the State of Colorado, which body shall determine whether or not discrimination has been shown, and as respects any representative deemed by the Commission to have been unfairly dealt with, the company shall make such reparation as the State of Colorado Industrial Commission may deem just.

IV

SOCIAL AND INDUSTRIAL BETTERMENT

1. Executive supervision.

The president's executive assistant, in addition to other duties, shall, on behalf of the president, supervise the administration of the company's policies respecting social and industrial betterment.

2. Co-operation of president's executive assistant with joint committees in carrying out policies of social and industrial betterment.

In the discharge of his duties, the president's executive assistant shall from time to time confer with the several Joint Committees, on Industrial

Cooperation and Conciliation, on Safety and Accidents, on Sanitation, Health and Housing, and on Education and Recreation, appointed at the annual joint conferences, as to improvements or changes likely to be of mutual advantage to the company and its employes. Members of the several joint committees shall be at liberty to communicate at any time with the president's executive assistant with respect to any matters under their observation or brought to their attention by employes or officials of the company, which they believe should be looked into or changed. As far as may be possible, employes should be made to feel that the president's executive assistant will welcome conferences with members of the several joint committees on matters of concern to the employes, whenever such matters have a direct bearing on the industrial, social, and moral well-being of employes and their families or the communities in which they reside.

3. Advisory Board on Social and Industrial Betterment.

In addition to consulting, from time to time, the several joint committees or their individual members, the president's executive assistant shall be the chairman of a permanent Advisory Board on Social and Industrial Betterment, to which may be referred questions of policy respecting social and industrial betterment and related matters requiring executive action.

4. Members of Advisory Board.

The Advisory Board on Social and Industrial Betterment shall be composed of such of the company's officers as the president may designate.

5. Regular and special meetings of Advisory Board.

The Advisory Board shall meet at least once in every six months, and may convene for special meetings upon the call of the chairman whenever he may deem a special meeting advisable.

6. Powers and duties of the Advisory Board.

The Advisory Board shall have power to consider all matters referred to it by the chairman, or any of its members, or by any committee or organization directly or indirectly connected with the company, and may make such recommendations to the president as in its opinion seem to be expedient and in the interest of the company and its employes.

7. Supervision of community needs by president's executive assistant.

The president's executive assistant shall also exercise a general supervision over the sanitary, medical, educational, religious, social, and other like needs of the different industrial communities, with a view of seeing that such needs are suitably and adequately provided for, and the several activities pertaining thereto harmoniously conducted.

8. Method of carrying out improvements.

Improvements respecting social and industrial betterment shall, after approval by the president, be carried out through the regular company organization.

9. Hospitals and doctors.

In camps where arrangements for doctors and hospitals have already been made and are satisfactory, such arrangements shall continue.

In making any new arrangement for a doctor, the employes' representatives in the camps concerned, the president's executive assistant, and the chief medical officer shall select a doctor, and enter into an agreement with him which shall be signed by all four parties.

10. Company periodical.

The company shall publish, under the direction of the president's executive assistant, a periodical which shall be a means of communication between the management, the employes and the public, concerning the policies and activities of the company. This periodical shall be used as a means of coordinating, harmonizing, and furthering the social and industrial betterment work, and of informing employes of the personnel and proceedings of conferences, boards and committees in which they are interested. It shall record events pertaining to social and industrial activities, and be a medium for making announcements with reference to the same, and for diffusing information of mutual interest to the company and its employes.

11. Cost of administering plan of representation and of furthering social and industrial betterment policies.

The promotion of harmony and goodwill between the company and its employes and the furtherance of the well-being of employes and their families and the communities in which they reside being essential to the successful operation of the company's industries in an enlightened and profitable manner, the expenses necessarily incident to the carrying out of the social and industrial betterment policies herein described, and the plan of representation, joint conferences and joint meetings herein set forth, including the payment of traveling expenses of employes' representatives when attending joint conferences and annual joint meetings, and their reimbursement for the working time necessarily lost in so doing, shall be borne by the company. But nothing herein shall preclude employes of the company from making such payment to their representatives in consideration of services rendered on their behalf as they themselves may voluntarily desire and agree to make.

MEMORANDUM OF AGREEMENT

Respecting Employment, Living and Working Conditions Between the Colorado Fuel and Iron Company and its Employes in the Coal Mines and Coke Oven Plants in the State of Colorado

October, 1915, to January, 1918

It is mutually understood and agreed that in addition to the rights and privileges guaranteed the employes and the company, in the industrial representation plan herewith, the following stipulations respecting employment, living and working conditions shall govern the parties hereto from the date of their signatures hereon until January 1, 1918, and shall continue thereafter subject to revision upon ninety days' notice by either of the parties:

I
RENT OF DWELLINGS, LIGHT AND WATER

The charge to employes for dwellings without bath shall not exceed two dollars per room per month.

The present uniform charge of forty cents per electric light per month, with free light on porches, shall not be increased.

There shall be no charge for domestic water, except in cases where the company is obliged to purchase the same; in such cases the charges shall be substantially cost to the company.

II
PRICES OF POWDER AND DOMESTIC COAL

The rates to be charged employes for powder and domestic coal shall be substantially their cost to the company.

III
FENCING OF EMPLOYES' HOMES AND GARBAGE REMOVAL

To encourage employes to cultivate flower and vegetable gardens, the company agrees to fence, free of charge, each house lot owned by it.

The company will continue its practice of removing garbage free of charge.

IV
BATH AND CLUB HOUSES

As the need becomes manifest, the company will continue its present policy of providing, as rapidly as possible, suitable bath houses and social centers in the nature of club houses, for its employes at the several mining camps.

V
HOURS OF LABOR

Eight hours shall constitute a day's work for all employes working underground and in coke ovens. This shall mean eight hours exclusive of the noon hour and the time required to go and come from the mine opening to the place of employment.

Nine hours shall constitute a day's work for all other outside labor, except firemen and engineers.

VI
SEMI-MONTHLY PAYMENT OF WAGES

All employes shall be paid semi-monthly by check.

No deductions shall be made from earnings, except where authorized by employes.

APPENDIX I

VII
WAGE SCHEDULE AND WORKING CONDITIONS

No change affecting conditions of employment with respect to wages or hours shall be made without first giving thirty days' notice, as provided by statute.

The schedule of wages and the working conditions now in force in the several districts shall continue without reduction, but if, prior to January 1, 1918, a general increase shall be granted in competitive districts in which the company does not conduct operations, a proportional increase shall be made. For this purpose a joint meeting of the miners' representatives and proper officers of the company shall be called within thirty days after the increase in competitive districts is effective to discuss and determine an equitable method for fixing the new scale in the districts affected.

We hereby certify that the Plan of Representation and Agreement as set forth were discussed and unanimously adopted at a joint conference of the officers of The Colorado Fuel and Iron Company and the representatives of its employes, held at Pueblo, to-day, Saturday, October 2, 1915, and referred by the conference for approval to the Board of Directors of the company and to the company's employes at the several mining camps, on the understanding that the same should be voted upon by secret ballot, and if adopted by the Board of Directors on the one hand and a majority of the company's employes on the other, should become binding upon the parties thereto.

(Signed) C. J. HICKS,
Representing the Company.

(Signed) W. E. SKIDMORE,
Representing the Employes.
(Joint Secretaries of the Conference.)
Pueblo, Colo., October 2, 1915.

I hereby certify that the Plan of Representation and Agreement referred for approval to the Board of Directors of the company by the joint conference of the officers of The Colorado Fuel and Iron Company and the representatives of its employes, held at Pueblo on Saturday, October 2, 1915, was to-day duly considered by the Board of Directors and unanimously adopted.

(Signed) J. A. WRITER, *Secretary of the Board of Directors of The Colorado Fuel and Iron Company.*

We hereby certify that the Plan of Representation and Agreement referred for approval to the company's employes at the several mining camps by the joint conference of officers of The Colorado Fuel and Iron Company and the representatives of its employes, held at Pueblo on Saturday, October 2, 1915, was voted upon by secret ballot on Monday, Tuesday, Wednesday and Thursday, October 4, 5, 6, 7 and on Monday, October 25, 1915, and that having examined the official returns duly certified to by the tellers elected to take charge of the vote at the several camps, we find that the total number of votes cast was 2846, of which number 2404, or 84.47 per cent. of the total votes cast, were in favor of, and 442, or 15.53 per cent., were against the proposed plan and agreement.

(Signed) C. J. HICKS,
Representing the Company.

(Signed) W. E. SKIDMORE,
Representing the Employes.
Sopris, Colo., October 26, 1915.

APPENDIX 2: EMPLOYEE REPRESENTATIVES AT CF&I COAL MINES, 1915–1928, AND EMPLOYEE REPRESENTATIVES AT MINNEQUA WORKS, 1916–1928

Table A2.1. Employee representatives at CF&I coal mines, 1915–1928

1915	Nationality	Yr. Hired [or First Year as Rep.]	Job When Elected
Joe August	*	*	*
Attilio Bevaqua	*	*	*
Vino Blondo	*	*	*
William Brown	*	*	*
Louis Cerruti	*	*	*
Archie Dennison	American	*	rock man
D. Garcia	Mexican American	1913	miner
Nick Keseric	Austrian	1912	*
George A. Lewis	Welsh	*	*
Joe Lucero	American	*	miner
Luis Lusik	Austrian	1912	miner
Richard Madonna	Italian	1915	miner
Charles Mitchell	*	*	*
Dan Morelli	Italian	1914	miner
Joe Nacearatto	Italian	1913	*
Charles Ossalo	*	*	*
Walter Patrick	Scotch	1913	*

continued on next page

The primary sources for material in this table are the personnel cards in the CF&I Archives, supplemented by information from the *CF&I Industrial Bulletin* and the *CF&I Blast*. Many cards are missing or are so disorganized that my research assistants and I have been unable to find them. Missing information is marked by a *. Dates in brackets represent the first year an employee representative served in that capacity rather than his first year with the firm. For mor information, see Chapter 4, note 26.

APPENDIX 2

1915—continued	Nationality	Yr. Hired [or First Year as Rep.]	Job When Elected
D. R. Patterson	*	*	*
John Pernich	*	*	*
Joe Poleski	*	*	*
Byron Richards	*	*	*
Mike Ritz	Slav	1901	driver
H. J. Shoupe	American	1914	tipple boss
William Skidmore	American	1913	engineer
F. E. Songer	*	*	*
S. P. Thomas	*	*	*
Fred Turra	Italian	1914	miner
James R. Walton	*	*	*
Joe Ward	American	*	*
R. G. Wiley	*	*	*

1916	Nationality	Yr. Hired [or First Year as Rep.]	Job When Elected
O. E. Anderson	*	*	*
Frank Angelini	Austrian	1911	miner
Attilio Bevaqua	Italian	1911	miner
Tony Bole	*	*	*
Oscar Brown	*	*	*
F. E. Campbell	American	1912	loader
John Chapman	*	*	*
Archie Dennison	American	[1915]	rock man
D. Garcia	*	[1915]	*
E. G. Gladstone	*	1915	carpenter
Vince Gloriocia	Italian	*	miner
J. R. Hale	American	1911	*
John Jones	*	*	*
Isaiah Kennedy	*	*	*
Frank Leonetti	Italian	1911	*
George A. Lewis	Welsh	*	*
Dominick Masero	Italian	*	*
M. Melendrez	*	*	*
Charles Mitchell	*	[1915]	*
Eugene Montoya	Mexican	*	miner
Mike Morris	Irish	*	miner
E. H. Myers	American	*	*
Joe Naccaratto	Italian	1913	*
Charles Ossola	*	[1915]	*
Tony Ossola	*	*	*
Juan Pacheco	Mexican	*	*
John Pernich	*	[1915]	*

continued on next page

APPENDIX 2

1916—continued	Nationality	Yr. Hired [or First Year as Rep.]	Job When Elected
Marco Peters	*	*	*
Frank Phillips	*	*	loader
George Price	*	*	*
Byron Richards	*	[1915]	*
Charles Robinson	*	*	*
George Scollick	Scotch	1893	miner
Matt Skala	*	*	*
William Skidmore	American	1913	engineer
Joe Toth	*	*	*
John Udovich	*	1913	track layer
Joe Ward	American	*	*
R. G. Wiley	*	[1915]	*

1917	Nationality	Yr. Hired [or First Year as Rep.]	Job When Elected
O. E. Anderson	*	[1916]	*
Joe Arnott	*	1915	*
C. Battisti	*	*	*
Anton Bole	*	[1916]	*
J. J. Burns	American	1917	rock man
Tom Chowan	American	*	*
Sam Ciccone	Italian	1916	watchman
David Davies	*	*	*
John Deldosso	Italian	1899	*
Archie Dennison	American	[1915]	rock man
Cesidio De Santis	Italian	1916	miner
Jake Gallegos	Mexican	1916	rock man
D. Garcia	Mexican American	1913	miner
William C. Gilbert	*	*	*
E. G. Gladstone	*	1915	carpenter
Vince Glorioso	Italian	[1916]	miner
Leo Griego	*	*	*
Joe Immordino	Italian	*	miner
Joseph Jones	*	*	*
Martin Kauser	German	1916	miner
C. J. Keller	*	*	*
Frank Leonetti	Italian	1911	*
Robt. Llewellyn	*	1916	miner
Pete Macari	*	*	*
William Manley	*	*	*
William McShane	American	1916	*
Charles Mitchell	*	[1915]	*
Eugene Montoya	Mexican	[1916]	miner

continued on next page

APPENDIX 2

1917—continued	Nationality	Yr. Hired [or First Year as Rep.]	Job When Elected
Juan Pacheco	Mexican	[1916]	*
Walter Patrick	*	*	*
F. L. Patterson	American	1912	carpenter
Will Pavell	Bulgarian	1917	*
Nick Rampon	*	*	*
William Richards	American	1917	miner
Alex Rubin	American	1913	*
George Scollick	Scotch	1893	miner
John Shaw	*	*	*
C. N. Stark	American	1913	*
Harry Stewart	American	1910	*
S. T. Thomas	*	*	*
W. R. Thompson	African American	1913	miner
Tim Valdez	Mexican	1913	watchman
S. J. Warrick	*	*	*
Ellis Williams	*	*	*
James Wilson	*	*	*

1918	Nationality	Yr. Hired [or First Year as Rep.]	Job When Elected
Lou M. Allen	American	1916	*
O. E. Anderson	*	[1916]	*
George Argeros	Greek	1916	mucker
Joe Arnott	*	1915	*
Joe August	*	*	*
Thomas Blyth	*	*	*
P. G. Cameron	American	1911	fire boss
Nick Cerjancic	*	*	*
Henry Charters	American	1916	lamp man
M. Cummings	American	1915	miner
Jasper Dambrosi	Italian	1914	miner
David Davies	*	[1917]	*
John Deldosso	Italian	1899	*
Archie Dennison	American	[1915]	rock man
T. M. Dennison	*	*	miner
David Evans	*	*	*
Joseph Fanelli	Italian	1915	hand driller
Virginia Fantin	Italian	1914	*
James Frem	*	*	*
D. Garcia	Mexican American	1913	miner
John Gardner	*	*	*
W. C. Gieser	German	*	miner
E. G. Gladstone	*	1915	carpenter

continued on next page

APPENDIX 2

1918—continued	Nationality	Yr. Hired [or First Year as Rep.]	Job When Elected
John Griego	*	*	*
Leo Griego	*	[1917]	*
Candido Guadagnoli	Italian	1910	miner
James Hay	Scottish	1909	miner
Oscar Johnson	American	*	*
C. J. Keller	*	[1917]	*
C. E. Kennedy	American	1915	tipple boss
Joe Levi	*	*	*
Regino Lobato	*	*	*
Guadalupe Lucero	Mexican	1914	miner
Pete Macari	*	[1917]	*
James McGowan	American	1911	clerk
Eugene Montoya	Mexican	[1916]	miner
Russell Patterson	American	1916	machine helper
Jesse Penny	*	*	*
Reese T. Roberts	*	*	*
Lee Santisteban	*	*	*
Mike Scarvarda	Italian	1912	miner
George Scollick	*	[1916]	*
S. H. Short	American	1916	miner
C. N. Stark	American	1913	*
John Surisky	*	*	*
John Tafoya	*	*	*
Thomas Tucker	*	*	*
Tim Valdez	Mexican	1913	watchman
Joe Wilgolz	*	*	*
Dick Young	*	*	*

1919	Nationality	Yr. Hired [or First Year as Rep.]	Job When Elected
O. E. Anderson	*	[1916]	*
Joe August	*	[1915]	*
Thomas Blyth	*	[1918]	*
Roy Chavez	Mexican American	*	pipe man
D. H. Davis	*	*	*
John Deldosso	Italian	1899	*
Tom M. Dennison	*	[1918]	miner
Tony Di Piazzia	*	*	*
William Dow	Negro	1908	miner
Dave Evans	*	[1918]	*
Joseph Fanelli	Italian	1915	hand driller
D. Garcia	Mexican American	1913	miner
Angelo Gargaro	Italian	1915	laborer

continued on next page

APPENDIX 2

1919—continued	Nationality	Yr. Hired [or First Year as Rep.]	Job When Elected
Carl Gates	American	1914	*
James Gay	English	*	miner
E. G. Gladstone	*	1915	carpenter
Adonagos Gomez	*	*	*
Ed. Gordon	French	*	laborer
Leo Griego	*	[1917]	*
R. B. Guerrero	Mexican	*	*
John Haddow	*	*	*
David L. Hansen	*	*	*
John Horsman	*	*	*
E. H. Jenkins	American	1917	*
Edward Johnson	*	*	*
Felix Lopez	*	*	*
Pete Macari	*	[1917]	*
Tony Marion	*	*	*
Robert A. Marshall	*	*	*
John Merritt	*	*	*
Juan E. Mestas	*	*	*
Julian Muniez	*	*	*
Juan Pacheco	Mexican American	*	motorman
Russell Patterson	*	*	*
W. W. Penny	*	*	*
Louis Rino	Italian	*	*
Mike Redgich	*	*	*
Tom Roberts	*	1911	night boss
Fred Salvatore	*	*	*
Mike Scarvarda	Italian	1912	miner
Louis Shane	*	*	*
Antone Sneller	*	*	*
Thomas Tucker	*	[1918]	*
Anton Turkovich	Austrian	*	miner
Elbert Tyson	American	1916	driver
Tim Valdez	Mexican	1913	watchman
Joe Weilgosz	*	*	*
Tom Wilson	American	1910	laborer

1920	Nationality	Yr. Hired [or First Year as Rep.]	Job When Elected
O. E. Anderson	*	[1916]	*
William Anderson	*	*	*
Thomas Blyth	*	[1918]	*
Charles Brown	*	*	*
Jas. M. Brown	American	1912	*

continued on next page

APPENDIX 2

1920—continued	Nationality	Yr. Hired [or First Year as Rep.]	Job When Elected
Fred Brunelli	Italian	1911	miner
Nic Cerjanic	*	1913	loader
Roy Chavez	Mexican American	[1919]	pipe man
M. Cordova	*	*	*
Solomon Daminquez	*	*	*
Tom Davies	*	*	*
John Deal	American	*	*
John Deldosso	Italian	1899	*
Archie Dennison	American	[1915]	rock man
Dave Evans	*	[1918]	*
J. C. Fernandez	Mexican American	[1919]	fire boss
D. Garbiso	Mexican	1917	laborer
Angelo Gargaro	Italian	1915	laborer
Wm. C. Gieser	German	1911	miner
E. G. Gladstone	*	1915	carpenter
Juan Griego	*	*	*
G. B. Gunico	*	*	*
Milton Hobbs	American	1911	motorman
Ed Jenkins	American	1917	*
J. W. Landsam	*	*	*
H. A. Larsen	*	1919	weigh boss
J. A. Leyba	*	*	*
Frank Luna	Mexican	1914	pumper
Eugene Montoya	Mexican	[1916]	miner
John Morritt	*	*	*
Z. R. Orozco	Mexican	1914	loader
W. W. Penny	*	[1919]	*
Juan Pacheco	Mexican American	[1919]	motorman
Celedon Salazar	Mexican	1915	driver
Herbert Shaw	*	*	*
Steve Simones	*	*	*
L. O. Smith	*	*	*
P. J. Smith	*	*	*
James Tedesko	*	*	*
J. T. Thompson	*	*	*
Tim Valdez	Mexican	1913	watchman
John Volk	Austrian	1916	*

1921	Nationality	Yr. Hired [or First Year as Rep.]	Job When Elected
Ben Abeyta	Mexican	*	miner
O. E. Anderson	*	[1916]	*
Wm. Anderson	*	[1920]	*

continued on next page

APPENDIX 2

1921—continued	Nationality	Yr. Hired [or First Year as Rep.]	Job When Elected
Sam Andrews	American	*	pit boss
Mike Archuleta	Mexican	1918	blacksmith
Joe Barber	Italian	1915	miner
Max Bargas	*	*	*
Tony Bartlo	Italian	1907	miner
Thomas Blyth	*	[1918]	*
Joe Borella	*	*	*
Fred Brunelli	Italian	1911	miner
Nick Buffalo	Italian	*	*
Frank Champion	Dutch	1904	driver
J. B. Cunico	Italian	1915	*
Pete Detro	*	*	*
Salvatore Di Santis	*	*	*
F. C. Dyer	American	1919	machine runner
J. C. Fernandez	Mexican American	[1919]	fire boss
Philip Ferkovitch	Austrian	1916	loader
John Fox	American	1919	miner
E. G. Gladstone	*	1915	carpenter
Glen Godfrey	American	*	*
Candido Guadagnoli	*	*	*
Mose Hanson	American	1913	*
James Hay	Scotch	1915	miner
Ed Jenkins	American	1917	*
J. C. Keller	*	*	*
Frank Luna	Mexican	1914	pumper
J. D. Martinez	*	*	*
Jose L. Martinez	*	*	*
Bob Mathison	*	*	*
D. J. McCarthy	*	*	*
Thos. McNeill	*	1916	rock man
John Meek	*	*	*
John Merritt	*	*	machine runner
Frank Morelli	Italian	1910	miner
J. A. Overand	Mexican American	1911	mine pumper
Tom Palmer	Canadian	1915	miner
Charles Peete	American	1918	weigh man
Dan Quantrini	*	*	*
Celedon Salazar	Mexican American	1915	driver
George Scollick	Scotch	1893	miner
Glen Smith	American	1913	miner
L. O. Smith	*	[1920]	*
J. T. Thompson	*	[1920]	*
Gust Tomlich	*	*	*
Ben Valdez	Mexican American	[1920]	driver boss

continued on next page

APPENDIX 2

1921—continued	Nationality	Yr. Hired [or First Year as Rep.]	Job When Elected
Roy Warrick	American	1916	*
Jack Williams	*	*	*
Harry Wright	American	1915	*

1922	Nationality	Yr. Hired [or First Year as Rep.]	Job When Elected
James Arnott	American	1916	machinist
Mike Bear	Croatian	1912	driver
Tom Bodycomb	Welsh	1906	driver boss
A. J. Carlson	*	*	*
Roy Chavez	Mexican American	[1919]	pipe man
C. W. Crabtree	*	*	*
Sidney Dawe	American	1913	driver
Tom M. Dennison	*	[1918]	miner
J. C. Fernandez	Mexican American	[1919]	fire boss
John Fox	American	1919	miner
C. J. Keller	American	1916	motor man
C. A. Knapp	American	1921	pumpman
E. S. Gibson	*	*	*
Wm. C. Gilbert	English	1915	miner
Candido Guadagnoli	*	[1921]	miner
J. T. Harvey	*	*	*
Ed H. Jenkins	American	1913	driver
Thomas John	Welsh	*	miner
E. W. Johnson	Irish	1911	contractor
Frank Luna	Mexican	1914	pumper
Harry McCluskey	Scotch	1922	night boss
John McKeown	American	1921	lineman
John Merritt	*	[1921]	machine runner
W. E. Mitchell	American	1919	timberman
Mike Morelli	Italian	1920	lamp man
Russell Patterson	American	1916	machinist helper
Thos. Payne	American	1911	miner
Charles Peete	American	1918	weigh man
W. C. Raish	American	1918	loader
Celedon Salazar	Mexican American	1915	driver
Matthew Scanlon	*	*	boilermaker
A. J. Stapleton	American	1914	miner
M. C. Umphress	American	1920	machinist helper
Jack Williams	*	[1921]	*

continued on next page

APPENDIX 2

1923	Nationality	Yr. Hired [or First Year as Rep.]	Job When Elected
Wm. Aitken	*	*	machine runner
Victor Arnoldi	Italian	1906	miner
Tom Bodycomb	Welsh	1906	driver boss
Owen M. Burns	*	*	motorman
A. J. Carlson	*	[1922]	machinist
R. E. Chambers	*	*	outside foreman
Roy Chavez	Mexican American	[1919]	pipe man
W. T. Cook	*	1917	miner
Ed. Cordero	*	*	miner
S. A. Cruz	*	*	rope rider
W. F. Daugherty	American	1912	driver
Moses Davies	American	1915	miner
Archie Dennison	American	[1915]	fire boss
Thos. M. Dennison	*	[1918]	miner
J. A. Deus	Spanish American	1918	weigh boss
Sam Di Giacomo	*	*	miner
Thos. Dowd	*	*	section foreman
J. C. Fernandez	Mexican American	[1919]	fireboss
Wm. C. Gieser	German	1911	miner
Wm. C. Gilbert	English	1915	miner
John Goldsby	American	1918	roller man
Candido Guadagnoli	*	[1921]	miner
David Hunden	American	1899	track layer
Thomas John	American	*	miner
H. L. Johnson	American	*	timberman
C. J. Keller	American	1916	motorman
Edmund Kitto	American	1917	lamp lighter
Chas. H. Lee	American	1918	hoist
A. L. Lumsden	*	*	pumpman
Abel Martinez	*	*	machine runner
Candido Martinez	Mexican American	1917	rock man
L. A. Mayer	*	*	*
John Merritt	*	[1921]	machine runner
J. E. Mestas	*	*	machine runner
J. A. Overand	Mexican	1911	mine pumper
Tom Palmer	Canadian	1915	miner
Charles Peete	American	1918	weigh man
Wm. Price	American	1921	driver
Celedon Salazar	Mexican American	1915	running motor
Pete Salvatorte	Italian	1908	shot firer
Matthew Scanlon	*	[1922]	boilermaker
Frank Sisneros	Mexican	1916	coke ovens
I. W. Wiles	*	*	pumpman

continued on next page

APPENDIX 2

1924	Nationality	Yr. Hired [or First Year as Rep.]	Job When Elected
Archie Allison	Scotch	*	lamp lighter
James Arnott	American	*	driver
Joseph Arnott	American	*	*
Mike Bear	Croatian	1912	miner
R. E. Bludworth	American	*	rope rider
O. M. Burns	American	[1922]	pumper
Roy Chavez	Mexican American	[1919]	fire boss
W. C. Cook	American	*	rock man
Archie Dennison	American	[1915]	motorman
J. A. Deus	Mexican American	1918	trip rider
Walter Dougherty	American	1912	driver
J. C. Fernandez	Mexican American	[1919]	hoist man
John Fox	American	*	coal contractor
Dave Garbiso	Mexican American	*	miner
Diego Garcia	Mexican American	*	miner
Malvern Godfrey	American	*	miner
John Goldsby	American	1918	miner
William H. Gray	American	*	machine runner
Thomas John	American	[1922]	machine runner
H. L. Johnson	American	*	miner
Edmund Kitto	American	1917	rock man
C. A. Knapp	American	*	miner
C. H. Lee	American	1918	motorman
Candido Martinez	Mexican American	1917	shot firer
Fred Martinez	Mexican	*	track layer
John Maruelli	Italian	*	machinist
William McShane	American	*	breaker boss
Mike Morelli	Italian	*	machine runner
Juan Pacheco	Mexican American	[1919]	trip rustler
Tibo Paneda	Mexican American	*	miner
Charles Peete	American	1918	miner
William Richards	American	*	roller man
Charles Ross	American	*	driver boss
Pete Salvatorte	Italian	1908	fire boss
Alex Samora	Mexican American	*	rock man
Roy Simpleman	Mexican American	*	fire boss
Frank Sisneros	Mexican American	*	machinist
J. L. Smith	American	*	pipe man
H. E. Thielbar	American	*	weigh boss
Anton Turkovich	Austrian	[1919]	charger
H. W. Vahldick	American	*	miner
Ben Valdez	Mexican American	[1920]	trip rider
Gwilym Williams	Welsh	*	miner

continued on next page

APPENDIX 2

1925	Nationality	Yr. Hired [or First Year as Rep.]	Job When Elected
Archie Allison	Scotch	[1924]	lamp lighter
James Arnott	American	[1924]	driver
Peter Campbell	*	*	*
Roy Chavez	Mexican American	[1919]	fire boss
W. T. Cook	*	1917	miner
Archie Dennison	American	[1915]	motorman
Thos. M. Dennison	*	[1918]	miner
John Deldosso	Italian	1899	*
Cesisio DeSantis	Italian	1900	miner
Walter Dougherty	American	1912	*
William Ellis	American	1910	miner
John Falgien	Italian	1906	miner
J. C. Fernandez	Mexican American	[1919]	hoist man
Frank Fidino	Italian	1921	miner
Dave Garbiso	Mexican American	*	miner
Glen Godfrey	American	*	*
John Goldsby	American	1918	miner
B. S. Goss	American	1911	machine runner
R. E. Jenkins	*	*	*
Edward Johnson	*	*	*
Edmund Kitto	American	1917	rock man
Charles Lee	American	1918	motorman
Candido Martinez	Mexican American	1917	shot firer
J. D. Martinez	*	*	*
J. E. Mestas	*	[1923]	machine runner
Elipio Mondragon	Mexican American	1911	electrician helper
William Murphy	American	1920	miner
Juan Pacheco	Mexican American	[1919]	trip rustler
Tibo Paneda	Mexican American	[1924]	miner
Henry Patterson	American	1900	pumpman
Joe Peduzzi	Italian	1916	fire boss
Charles Peete	American	1918	weigh man
William Richards	American	[1924]	roller man
B. B. Richardson	American	1900	fire boss
Joe Shain Sr.	Austrian	1921	miner
Frank Short	Italian	1916	miner
Roy Simpleman	Mexican American	[1924]	fire boss
J. L. Smith	American	[1924]	pipe man
H. E. Thielbar	American	*	weigh boss
Ben Valdez	Mexican American	[1920]	trip rider
Jack Williams	*	[1921]	*
Adam Young	*	*	*

continued on next page

APPENDIX 2

1926	Nationality	Yr. Hired [or First Year as Rep.]	Job When Elected
Frank Anaya	*	*	*
David Arthur	American	1885	miner
Epiano Atencio	Mexican American	1921	miner
Frank Birge	Mexican American	1904	driver
G. B. Botsford	American	1916	*
Albert Brown	African American	1916	driver-miner
Isidor Bueno	Mexican	*	*
Peter Campbell	*	[1925]	*
A. J. Carlson	*	*	*
John Concillia Jr.	*	*	*
Andy Conder	*	*	*
Moses Davies	American	1915	miner
John Deldosso	Italian	1899	*
Frank Donati	Italian	1903	*
George Drumright	American	*	loader
John Embleton	American	1900	*
J. C. Fernandez	Mexican American	[1919]	hoist man
Glen Godfrey	American	[1925]	*
Thomas John	American	*	miner
Edward Johnson	*	[1925]	*
Tony Krizmanich	Austrian	1915	miner
Alfred Laiminger	*	*	*
J. D. Martinez	*	[1925]	*
Salvator Martinez	Mexican	1922	machine miner
Frank Nacaratto	*	*	*
Joe Oberster	*	*	*
E. A. Ortiz	Mexican	1907	loader
Juan Pacheco	Mexican American	[1919]	trip rustler
Tibo Paneda	Mexican American	[1924]	miner
Joe Peduzzi	Italian	1916	fire boss
Charles Peete	American	1918	weigh man
Claud Philpott	American	1923	miner
Victor Ribal	Mexican	1913	miner
Byron Richards	*	[1915]	*
B. B. Richardson	American	1900	fire boss
J. L. Smith	American	[1924]	pipe man
Ben Valdez	Mexican American	[1920]	trip rider
Adam Young	*	[1925]	*
John Zubal	*	*	*

continued on next page

APPENDIX 2

1927	Nationality	Yr. Hired [or First Year as Rep.]	Job When Elected
Frank Anaya	*	[1926]	*
David Arthur	American	1885	miner
Joe Barber	Italian	1909	miner
Frank Birge	Mexican American	1904	driver
A. J. Carlson	*	[1926]	*
John Concillia Jr.	*	[1926]	*
Andy Conder	*	*	*
John Deldosso	Italian	1899	*
Frank Donati	Italian	1903	*
Oliver Edwards	American	1925	*
Ben Gates	American	1907	blacksmith
Dan Jaramillo	Mexican American	1914	driver
Alfred Laiminger	*	[1926]	*
Frank Madonna	*	*	*
John Madonna	*	*	*
J. D. Martinez	*	[1925]	*
Anton Matkovich	*	*	*
Joe Micheli	Italian	1917	loader
Andrew Nicol	American	1905	*
E. A. Ortiz	Mexican	1907	loader
Juan Pacheco	*	[1926]	*
Tibo Paneda	Mexican American	[1924]	miner
Joe Peduzzi	Italian	1916	fire boss
Claud Philpott	American	1923	miner
Byron Richards	*	[1915]	*
Glen Smith	*	*	*
J. L. Smith	American	[1924]	pipe man
Thos. Templeman	*	*	*
Ben Valdez	Mexican American	[1920]	trip rider
S. J. Warrick	English	1896	pumper
T. L. Wilson	Scotch	1922	coal inspector
Adam Young	*	[1925]	*

1928	Nationality	Yr. Hired [or First Year as Rep.]	Job When Elected
David Arthur	American	1885	miner
Joe Barber	Italian	1909	miner
Frank Birge	Mexican American	1904	driver
Albert Brown	*	*	*
Nello C. Cunico	Italian	*	*
John Deldosso	Italian	1899	*
Walter Dougherty	American	1912	*

continued on next page

APPENDIX 2

1928—continued	Nationality	Yr. Hired [or First Year as Rep.]	Job When Elected
L. O. English	American	*	machine miner
Dan Jaramillo	*	*	*
Thomas John	American	*	miner
Wm. Komora	American	1912	miner
Alfred Laiminger	*	[1926]	*
Charles Lee	American	1918	motorman
Dan N. Lynch	*	*	*
Frank Madonna	*	[1927]	*
R. A. Marshall	*	*	*
J. D. Martinez	*	[1925]	*
Anton Matkovich	*	[1927]	*
Pat McKeown	American	1889	miner
Esquiel Mestas	Mexican	1908	*
Joe Micheli	Italian	1917	loader
Mike Morelli	Italian	[1924]	machine runner
E. A. Ortiz	Mexican	1907	loader
Tibo Paneda	Mexican American	[1924]	miner
Claud Philpott	American	1923	miner
Byron Richards	*	[1915]	*
Samuel Sanchez	Mexican	*	*
Mose Sena	Mexican American	*	loader
Glen Smith	*	[1927]	*
J. L. Smith	American	[1924]	pipe man
Wm. Stanley	American	*	miner
Thos. Templeman	*	[1927]	*
Nick Trujillo	*	*	*
Ben Valdez	Mexican American	[1920]	trip rider
Juan Vigil	*	*	*
J. W. Whisenaut	African American	1901	miner
Joe Wilkins	American	1907	*

APPENDIX 2

Table A2.2. Employee representatives at Minnequa Works, 1916–1928

1916	Nationality	Yr. Hired [or First Year as Rep.]	Job When Elected
Chas. B. Bartle	*	*	*
G. C. Bebout	*	1906	staple maker
H. J. Butzbach	American	1914	*
D. C. Cook	American	1914	electrician
Roy N. Doughty	*	1914	storekeeper
John Fahey	American	1903	labor
Ed. Floyd	American	*	labor
V. E. Johnson	*	*	drawer
Tony Juliano	*	*	*
Anton Kochevar	*	1904	helper
Wm. Lamb	*	1914	roller
A. H. Lee	American	1912	crane
H. O. Lemon	American	1913	operator
Fred Lobaugh	*	*	*
F. Jos. Loefller	*	*	*
Pedro Lopez	*	*	*
C. N. MacDonald	American	*	machinist
Dennis Mallon	*	1908	adjuster
Geo. W. McCray	American	1914	engineer
Thos. P. Reese	American	1906	nut maker
Leslie Sands	*	1898	turner
Gotlieb Schultz	*	*	*
B. J. Thomas	Scotch	1909	mason
Frank Van Dyke	*	1911	second helper
W. H. Walker	American	1914	skip hoist
Geo. J. White	American	1901	wire drawer

See Table A2.1 for details on the compilation of this chart.
* Information is not available.

1917	Nationality	Yr. Hired [or First Year as Rep.]	Job When Elected
Wilbur Barnes	American	1898	*
C. B. Bartle	*	[1916]	*
H. T. Bennett	*	*	furnace man
William Beurman	*	*	*
C. H. Bradley	American	1910	weigh man
A. J. Collins	American	1914	laborer
John Curran	*	*	*
C. T. Darnell	*	*	labor
Warren Densmore	American	1907	millwright
Andy J. Diamond	American	1914	roll turner

continued on next page

APPENDIX 2

1917—continued	Nationality	Yr. Hired [or First Year as Rep.]	Job When Elected
R. N. Doughty	*	1914	storekeeper
John Fahey	American	1903	labor
John T. Glover	American	1915	inspector
Hugh Goff	*	*	*
C. J. Grundy	American	1903	heater
C. H. Harpel	*	*	*
E. J. Hendrickson	American	1913	pump engineer
Robert Heustis	American	1901	first helper
Richard Inman	*	1903	millwright
V. E. Johnson	*	[1916]	drawer
Ira Kirkman	*	1914	melter
A. H. Lee	American	1912	crane
Fred Lobaugh	*	[1916]	*
Dennis Mallon	*	1908	adjuster
John McGann	*	1898	roller
Thomas McGee	*	*	*
William J. Morris	*	*	*
J. H. Raley	*	*	*
Thomas P. Reese	American	1906	nut maker
J. S. Rogers	*	*	*
Leslie Sands	*	1898	turner
Gotlieb Schultz	*	[1916]	*
Ralph Sears	American	1914	engineer
Charles S. Vore	*	*	*
George J. White	American	1901	wire drawer

1918	Nationality	Yr. Hired [or First Year as Rep.]	Job When Elected
Wilbur Barnes	American	1898	*
G. C. Bebout	*	*	*
William Beurman	*	[1917]	*
E. K. Boyd	American	1914	machinist
C. H. Bradley	American	1910	weigh man
James Brown	*	*	*
John Burkhart	*	*	*
James A. Comisky	*	*	*
Adolph A. Cunning	*	*	*
E. J. Daniels	*	1906	millwright helper
Warren Densmore	American	1907	millwright
Andy J. Diamond	American	1914	roll turner
Henry Fuerhardt	American	1915	filler
Hugh Goff	*	[1917]	*
John E. Gross	American	1914	machinist

continued on next page

APPENDIX 2

1918—continued	Nationality	Yr. Hired [or First Year as Rep.]	Job When Elected
C. J. Grundy	American	1903	heater
W. A. Haney	American	1914	labor
E. O. Harrington	*	*	*
V. E. Johnson	*	[1916]	*
J. B. Jones	*	*	*
T. Langford	*	*	*
Perry M. Lichty	*	*	*
F. Jos. Loeffler	*	[1916]	*
Elmer E. Lofgren	American	*	roll laborer
Chester Manville	American	*	helper
Mike Masar	*	*	*
Thomas McGee	*	[1917]	*
E. E. Morris	*	1912	helper
J. H. Raley	*	[1917]	*
John Redrauff	*	*	*
Gotlieb Schultz	*	[1916]	*
Ralph Sears	American	1914	engineer
Anton Spolar	*	*	*
George Strong	*	1904	groundskeeper
W. D. Turner	*	*	*
J. F. Tyner	*	1912	labor/helper
Charles S. Vore	*	[1917]	*

1919	Nationality	Yr. Hired [or First Year as Rep.]	Job When Elected
G. C. Bebout	*	[1916]	*
Thos. Bennett	*	*	*
Wm. Beurman	*	[1917]	*
Ed. Boilard	*	1914	guide tender
Henry W. Brannen	*	*	*
W. H. Corbett	American	1915	boilermaker
A. A. Cunning	*	[1918]	*
J. W. Dalton	*	*	*
E. J. Daniels	*	1906	millwright
Bert Davis	American	1911	handyman
M. C. Davis	*	1900	station engineer
W. M. Dayton	*	*	*
Warren Densmore	American	1907	millwright
Andy J. Diamond	American	1914	roll turner
Al Evans	*	*	*
Frank Ferguson	*	*	*
Henry Fuerhardt	American	1915	filler
John E. Gross	American	1914	machinist

continued on next page

APPENDIX 2

1919—continued	Nationality	Yr. Hired [or First Year as Rep.]	Job When Elected
C. J. Grundy	American	1903	heater
E. O. Harrington	*	[1918]	*
Wm. H. Howard	*	*	*
V. E. Johnson	*	[1916]	*
J. E. Jones	*	*	*
John Jones	*	*	*
Anton Kochevar	*	1904	helper
Tom Lally	American	*	crane man
Robt. Leithead	American	1914	machinist
L. C. Mather	*	1914	skip hoist
John J. McAdam	*	*	*
Levi B. Nutt	*	1907	first helper
J. H. Raley	*	[1917]	*
John Sabo	*	*	*
Gotlieb Schultz	*	[1916]	*
E. D. Shomaker	American	1906	pattern maker
Wolcott W. Smith	*	1912	armature
F. D. Spicer	American	1901	bale tie man
Ray Talbot	American	1913	electrical inspection helper
Guy Wilson	*	*	*
J. H. Windle	American	1909	*
E. Young	*	*	*

1920	Nationality	Yr. Hired [or First Year as Rep.]	Job When Elected
Al. Anderson	*	*	*
Clarence Banks	African American	1916	switch man
G. C. Bebout	*	[1916]	*
Thos. Bennett	*	*	*
C. H. Bradley	*	*	*
G. H. Bratton	American	1906	pattern maker
John Brennan	American	1914	notch
Fred Brown	*	*	*
Peter Collins	English	1902	blacksmith
C. E. Condit	*	*	*
A. A. Cunning	*	[1918]	*
M. C. Davis	*	1900	station engineer
Warren Densmore	American	1907	millwright
J. G. Emrick	American	1898	engineer
Ted Gager	*	1918	electrician
W. A. Gardiner	*	*	*
C. J. Grundy	American	1903	heater

continued on next page

APPENDIX 2

1920—continued	Nationality	Yr. Hired [or First Year as Rep.]	Job When Elected
Anton Guadagno	*	*	*
Wm. Hampton	*	*	*
Wm. Hogan	*	1914	barb stocker
M. C. Howard	*	*	*
J. E. Jones	*	*	*
Joe Jones	*	*	*
Chas. M. Lane	American	*	engineer
F. Jos. Loeffler	*	[1916]	*
Lester Lofgren	American	1917	laborer
George H. Love	American	*	labor
A. R. Lloyd	*	*	*
L. C. Mather	*	1914	skip hoist
R. L. McGuffin	*	1908	labor
John McMeekan	*	*	*
Frank M. Melton	*	*	carpenter
C. A. Miller	*	*	*
H. Moody	*	*	*
Wm. J. Morris	*	*	*
J. E. Munoz	*	*	*
August Nelson	*	1914	skull cracker
Wm. R. Perkins	American	1915	first oval
H. P. Pike	*	*	*
Wm. Preston	*	1913	labor
Chas. C. Ragland	*	*	watchman
Leslie Sands	*	1898	turner
W. D. Turner	American	1900	helper

1921	Nationality	Yr. Hired [or First Year as Rep.]	Job When Elected
Clarence Banks	African American	1916	switch man
George H. Bratton	American	1906	pattern maker
John Brennan	American	1914	first helper
W. A. Carver	American	*	Teamster
James Cochran	*	*	roll hand
Joe Cruz	*	1916	water tender
E. J. Daniels	American	1909	reel tender
M. C. Davis	*	1900	engineer
Warren Densmore	American	1907	millwright
Charles A. DeVore	*	*	heater
Richard Dolan	American	*	crane operator
S. Fowler	*	*	engineer
Anton Guadagno	*	[1920]	*
Joe Guadagno	*	*	*

continued on next page

APPENDIX 2

1921—continued	Nationality	Yr. Hired [or First Year as Rep.]	Job When Elected
J. S. Hammell	American	*	engineer
James Hight	*	*	*
James G. Hocking	*	1917	watchman
William Hogan	American	1914	rewinder
Richard Inman	American	*	millwright helper
Joe Jones	*	[1920]	roll hand
Geroge W. Joseph	*	*	layer out
E. J. Kittrell	*	1904	wire drawer
C. Lundstrom	*	1910	wiper
John McMeekan	*	[1920]	*
Rudolph Megler	*	*	*
Henry Metzler	*	*	*
William J. Morris	*	[1920]	brick mason
William O'Nell	*	*	cooper shop
Roy Palmer	*	1912	engineer
James Pierce	American	1908	moulder
Frank H. Pooler	*	1899	machinist
Frank Putnam	American	*	PN operator
Joseph Roach	*	*	machinist
W. J. Rogers	*	*	motor tender
John Sabo	*	[1919]	*
Dave Thomas	*	*	machinist
W. J. Thomas	*	*	electrician
J. F. Wheelan	*	*	hook on

1922	Nationality	Yr. Hired [or First Year as Rep.]	Job When Elected
A. C. Abercrombie	*	*	*
Louis Allito	Italian	[1922]	brick mason
Frank Artley	American	1911	*
Clarence Banks	African American	1916	switch man
John Brennan	American	1914	notch
Jose Castaneda	Mexican	1916	coal ash
M. C. Davis	*	1900	station engineer
Wm. Decco	*	*	*
Warren Densmore	American	1907	millwright
Richard Dolan	American	[1921]	crane operator
Wm. Dunlap	American	1915	brick layer
A. Gammon	*	*	*
Jas. H. Gentry	*	*	*
Joe Guadagno	*	[1921]	*
John Hackett	*	*	*

continued on next page

APPENDIX 2

1922—*continued*	*Nationality*	Yr. Hired *[or First Year as Rep.]*	*Job When Elected*
J. H. Hawse	*	*	*
Gus Helbeck	*	*	*
Clyde Helm	*	*	*
James Hight	*	*	*
Wm. H. Howard	*	[1919]	*
V. E. Johnson	*	[1916]	*
J. B. Jones	*	*	*
Geo. W. Joseph	*	[1921]	*
Jas. Kendall	*	*	*
F. Jos. Loeffler	*	[1916]	*
E. Lofgren	American	*	roll turner
S. H. Marone	*	*	*
H. Martillaro	*	*	*
John McMeekan	*	[1920]	*
Milan Milcich	*	*	*
Dave Morehart	*	*	*
R. S. Offerle	American	1912	blocker
Frank Pooler	*	1899	machinist
W. A. Pope	*	1899	heater
Frank Putnam	American	[1921]	*
J. L. Reid	American	1913	armature welder
Jos. Roach	*	[1921]	*
John Sabo	*	[1919]	*
Leslie Sands	*	1898	roll turner
Anton Schaffer	*	*	*
Joe. Schweary	*	1915	ash man
Chas. Shiner	*	*	*
Jos. Spinuzzi	*	1915	ring core
Chas. Sprinkle	American	1902	roll maker
Rhees Thomas	*	*	*
Roy Tiller	*	*	*
Manual Vega	*	*	*
L. A. Watson	*	1903	blacksmith
Geo. Webster	American	1912	help
Geo. White	American	1901	wire drawer
Sam Yelich	*	1918	cleaner

1923	*Nationality*	Yr. Hired *[or First Year as Rep.]*	*Job When Elected*
Louis Allito	Italian	[1922]	brick mason
A. G. Argabright	American	*	electrician
Clarence Banks	African American	1916	switch man

continued on next page

APPENDIX 2

1923—continued	Nationality	Yr. Hired [or First Year as Rep.]	Job When Elected
E. L. Benson	Swede	*	wire drawer
Fred S. Bigelow	*	1919	shipping
Paul Byrtus	*	*	*
W. F. Cole	*	*	catcher
Fred Daniels	American	1909	helper
M. C. Davis	*	1900	station engineer
Warren Densmore	American	1907	millwright
Andy J. Diamond	American	1914	roll turner
Richard Dolan	American	[1921]	crane operator
Anton Guadagno	*	[1920]	*
Gus Hellenbeck	German	*	toolmaker
Ben Henderson	*	1918	fireman
F. R. Hitchcock	American	1918	heater
A. R. Johnson	*	*	heater
J. B. Jones	*	[1922]	*
J. R. Jones	*	*	*
Bert Langford	American	1918	first helper
E. J. Markert	*	*	*
Mike Milcich	*	[1922]	*
L. M. Nelson	*	[1922]	*
R. S. Offerle	American	1912	blocker
D. O'Hare	*	*	*
James Pierce	American	1908	handyman
Frank Putnam	American	[1921]	*
Clyde Rider	*	1917	rougher
Geo. R. Rix	*	1915	roll turner
N. L. Robertson	American	*	machinist
John Sabo	Austrian	[1919]	straightener
J. T. Smith	American	*	nail operator
Charles Sprinkle	American	1902	roll maker
Lloyd Stevenson	*	[1922]	*
J. R. Van Dyke	*	1920	carpenter
Edward Walpole	American	*	pipe foundry
L. A. Watson	American	1903	blacksmith
George J. White	American	1901	wire drawer

1924	Nationality	Yr. Hired [or First Year as Rep.]	Job When Elected
Louis Allito	Italian	[1922]	brick mason
A. G. Argabright	American	*	electrician
Clarence Banks	African American	1916	switch man
E. L. Benson	Swede	[1923]	wire drawer

continued on next page

APPENDIX 2

1924—continued	Nationality	Yr. Hired [or First Year as Rep.]	Job When Elected
V. D. Brown	American	*	roller
Fred N. Coats	American	*	inspector
Walter F. Collins	American	*	adjuster
Charles E. Condit	American	*	machinist
Warren Densmore	American	1907	millwright
Andy J. Diamond	American	1914	roll turner
Richard Dolan	American	[1922]	crane operator
Gus Hellenbeck	German	[1923]	toolmaker
F. R. Hitchcock	American	1918	heater
J. Jaklovich	Austrian	*	cooper shop
Will Johnson	African American	*	gas house
Archie T. Jones	American	*	engineer
Mike Labella	Italian	*	trackman
Bert Langford	American	1918	furnace man
J. L. Lasater	American	*	watchman
Peter Lindstrom	Swede	*	millwright
William Linfoot	American	*	crane operator
C. N. MacDonald	American	[1916]	machinist
John Nolan	Irish	*	fireman
Levi Nutt	American	*	furnace man
R. S. Offerle	American	*	gallery whipper
James Pierce	American	1908	molder
Frank Putnam	American	[1921]	*
N. L. Robertson	American	[1923]	machinist
John Sabo	Austrian	[1919]	straightener
Leslie Sands	American	[1916]	roll turner
J. T. Smith	American	[1923]	nail operator
Charles Sprinkle	American	1902	roll maker
J. A. Stanko	Austrian	*	stocker
J. R. Van Dyke	American	*	washer man
Edward Walpole	American	*	pipe foundry
L. A. Watson	American	1903	blacksmith
George Wheelan	American	*	locomotive cranes
J. F. Wheelan	American	[1921]	torch operator
George J. White	American	1901	wire drawer

1925	Nationality	Yr. Hired [or First Year as Rep.]	Job When Elected
Ed Anderson	*	*	engineer
Clarence Banks	African American	1916	switch man
Jose Barelo	*	*	*
H. T. Bennett	*	[1917]	furnace man

continued on next page

APPENDIX 2

1925—continued	Nationality	Yr. Hired [or First Year as Rep.]	Job When Elected
C. H. Bradley	*	*	*
George Bravdica	*	*	adjuster
V. D. Brown	American	[1924]	roller
Albert Carillo	*	*	coke inspector
W. F. Cole	*	*	*
Charles E. Condit	American	[1924]	machinist
R. D. Davidson	*	*	*
M. C. Davis	*	1900	station engineer
Warren Densmore	American	1907	millwright
Andy J. Diamond	American	1914	roll turner
R. J. Dolan	American	[1922]	crane operator
Mike Filler	*	*	*
Frank W. Gotbehuet	*	*	*
Anton Guadagno	*	[1920]	*
Gus Hellenbeck	German	[1923]	toolmaker
Roy Imes	*	*	*
V. E. Johnson	*	[1916]	drawer
V. B. King	*	*	*
J. L. Lasater	American	[1924]	watchman
C. N. MacDonald	American	[1916]	machinist
George Martin	*	*	*
John McMeekan	*	[1920]	brick layer
Milan Milcich	*	[1925]	*
Dave Moreheart	*	[1922]	*
Arthur O'Donnell	*	*	*
John C. Perko	*	*	knife grinder
James Pierce	American	1908	handyman
Hiram W. Prior	*	*	roll turner
N. L. Robertson	American	[1923]	machinist
Sabina Sena	*	*	fireman
E. A. Sewell	*	*	*
Emmet Spees	*	*	*
Charles Sprinkle	American	1902	roll maker
Frank Taulie	*	*	*
Edward Walpole	American	*	pipe foundry
L. A. Watson	American	1903	blacksmith

1926	Nationality	Yr. Hired [or First Year as Rep.]	Job When Elected
Ed Anderson	*	[1925]	engineer
Clarence Banks	African American	1916	switch man
John A. Barnes	*	*	shipper

continued on next page

APPENDIX 2

1926—continued	Nationality	Yr. Hired [or First Year as Rep.]	Job When Elected
H. T. Bennett	*	[1917]	furnace man
Edwin Berg	*	*	charger
Geo. Bravdica	*	[1925]	adjuster
John I. Burdick	*	*	inspector
Wm. H. Buser	*	*	fireman
Albert Carillo	*	[1925]	coke inspector
Manley. H. Coffman	*	*	torch operator
Theodore E. Cook	*	*	stocker
R. D. Davidson	*	[1925]	*
Angelo DeSoto	*	*	rope splicer
Andy J. Diamond	American	1914	roll turner
Richard Dolan	American	[1922]	crane operator
Earl S. Ferguson	*	*	millwright
Mike Filler	*	[1925]	*
Robert Gorman	*	*	roll hand
Anton Guadagno	*	[1920]	trackman
John Hasling	*	*	machinist
Gus Hellenbeck	German	[1923]	toolmaker
A. R. Johnson	*	[1923]	heater
V. E. Johnson	*	[1925]	drawer
A. T. Jones	*	*	engineer
Anton Kochevar	*	1904	wire drawer
J. L. LaSater	*	*	watchman
Peter Lindstrom	Swede	[1924]	millwright
Harry Lobaugh	*	*	millwright
Roy Love	*	*	electrician
John McMeekan	*	[1920]	brick layer
Robert J. O'Brien	*	*	millwright
John C. Perko	*	[1925]	knife grinder
Hiram W. Prior	*	*	roll turner
Sabina Sena	*	[1925]	fireman
Charles Sprinkle	American	1902	roll maker
George Swearingen	*	*	locomotive crane operator
Roy T. Vaughn	*	*	straightener
Thomas J. Walsh	*	*	core maker
L. A. Watson	American	1903	blacksmith
L. A. Webber	*	*	electrician

continued on next page

APPENDIX 2

1927	Nationality	Yr. Hired [or First Year as Rep.]	Job When Elected
R. O. Andrews	*	*	*
Clarence Banks	African American	1916	switch man
Edwin Berg	*	*	charger
F. S. Bigelow	*	*	*
R. R. Bowman	*	*	*
M. H. Coffman	*	*	*
Theodore E. Cook	*	[1926]	stocker
R. D. Davidson	*	[1925]	*
M. C. Davis	*	1900	station engineer
Andy J. Diamond	American	1914	roll turner
Richard Dolan	American	[1922]	crane operator
Frank Fabian	*	*	*
Mike Filler	*	[1925]	*
Frank W. Gotbehuet	*	[1925]	*
Anton Guadagno	*	[1920]	trackman
Joe Guadagno	*	[1921]	*
John Hasling	*	[1926]	machinist
A. Hendrickson	*	*	*
Roy Imes	*	[1925]	*
Freeman H. Johnson	*	*	*
V. E. Johnson	*	[1925]	drawer
A. T. Jones	*	[1926]	engineer
Bert Langford	American	1918	furnace man
Peter Lindstrom	Swede	[1924]	millwright
Roy Love	*	[1926]	electrician
Preston Lovette	*	*	*
Ray McIntyre	*	*	*
John McMeekan	*	[1925]	brick layer
J. M. McQuay	*	*	*
Robert O'Brien	*	*	*
J. C. O'Donnell	*	*	*
John C. Perko	*	[1925]	knife grinder
W. P. Pitcock	*	*	*
B. L. Reynolds	*	*	*
Earl Rodgers	*	*	*
M. L. Rosky	*	*	*
Charles Sprinkle	American	1902	roll maker
Elias S. Strong	*	*	*
Roy T. Vaughn	*	[1926]	straightener
L. A. Webber	*	[1926]	electrician
J. W. Wheeler	*	*	*
George J. White	American	1901	wire drawer

continued on next page

APPENDIX 2

1928	Nationality	Yr. Hired [or First Year as Rep.]	Job When Elected
Lloyd Argabright	*	*	*
R. J. Beltzner	*	*	*
E. L. Benson	Swede	[1923]	wire drawer
Edwin Berg	*	[1927]	charger
J. Bowen	*	*	*
Albert Carillo	*	[1925]	coke inspector
Rawlin Carr	*	*	*
Theodore E. Cook	*	[1926]	stocker
M. C. Davis	*	1900	station engineer
Andy J. Diamond	American	1914	roll turner
Richard Dolan	American	[1922]	crane operator
Frank Fabian	*	*	*
Mike Filler	*	[1925]	*
Anton Guadagno	*	[1920]	trackman
Joe Guadagno	*	[1921]	*
John Hasling	*	[1926]	machinist
Hiram Hogan	*	*	*
V. E. Johnson	*	[1925]	drawer
Phillip Kaplan	*	*	*
Will F. Kaufman	*	*	*
Frank Kelly	*	*	*
Roy R. Levy	*	*	*
Peter Lindstrom	Swede	[1924]	millwright
Roy Love	*	[1926]	electrician
Mike Masar	*	*	*
Ray McIntyre	*	[1927]	*
John McMeekan	*	[1925]	brick layer
Robert O'Brien	*	[1927]	*
J. C. O'Donnell	*	[1927]	*
Clark L. Patch	*	*	*
John C. Perko	*	[1925]	knife grinder
Arthur Pike	*	*	*
W. P. Pitcock	*	[1927]	*
John Schmit	*	*	*
Steve Secora	*	*	*
Charles Sprinkle	American	1902	roll maker
Elias Strong	*	[1927]	*
Roy P. Templin	*	*	*
Edward Thomas	*	*	*
Denny O. Tipton	*	*	*
J. W. Wheeler	*	[1927]	*
J. H. Withers	*	*	*

NOTES

PREFACE

1. http://www.time.com/time/magazine/article/0,9171,745938,00.html?iid=chix-sphere, accessed February 10, 2009.

2. H. Lee Scamehorn, *Pioneer Steelmaker in the West: The Colorado Fuel and Iron Company 1872–1903* (Boulder: Pruett, 1976), v. Scamehorn covers the rest of the firm's history in *Mill & Mine: The CF&I in the Twentieth Century* (Lincoln: University of Nebraska Press, 1992), the source of the figures used in this paragraph.

3. The company was later bought by Oregon Steel, which operated the Pueblo facilities under the name Rocky Mountain Steel Mills, and is now owned by the Russian firm Evraz.

4. I revised and published this work as *Managing the Mills: Labor Policy in the American Steel Industry during the Nonunion Era* (Landover, MD: University Press of America, 2004).

5. The company's headquarters was in Denver until it moved to Pueblo during the early 1930s as a cost-saving measure.

6. In addition to the name of the inventor of a revolutionary steelmaking process, Bessemer is the name of the neighborhood in Pueblo in which the steel plant is located.

7. Colorado State University–Pueblo is co-owner, but the Bessemer Historical Society has operational control of the collection.

8. Boxes were often renumbered or re-filed over the course of my research for this book. However, as I note in the bibliographic essay at the end of the book, the name of the record group "Industrial Relations Materials" has consistently been the same. In this work I leave out box numbers in citations to this collection so as not to provide information that is outdated.

9. During the research process, I found many more published minutes from the 1928–1941 period in the National Archives outside Denver. These minutes are among the exhibits in the 1941 court case that ruled on the legality of the National Labor Relations Board's decision to invalidate the plan under the National Labor Relations Act as an illegal employee-dominated labor organization. *Colorado Fuel & Iron Corporation v. National Labor Relations Board*, 121 F.2d 165 (10th Cir., 1941).

10. William Manchester, *A Rockefeller Family Portrait: From John D. to Nelson* (Boston: Little, Brown, 1959), 112.

11. Mackenzie King Diary, October 24, 1916, Collections Canada, http://king.collectionscanada.ca. King heard the story from Rockefeller's aide Charles Heydt, who feared that if the story of Rockefeller Senior's CF&I stock strategy ever became public it would cause a huge scandal. Writing to his father in April 1917 to ask him to contribute to construction of a YMCA building in Pueblo, Rockefeller referred to "your former large interest" in the company, "which so far as the public generally knows still continues." See John D. Rockefeller Jr. to John D. Rockefeller Sr., April 13, 1917, Rockefeller Family Archives, Record Group III 2 C, Box 18, Folder 156, Rockefeller Archive Center, Sleepy Hollow, NY.

INTRODUCTION

1. *The Washington Post*, October 5, 1915.

2. Ernest Richmond Burton, *Employee Representation* (Baltimore: Williams & Wilkins, 1926), 19.

3. Daphne G. Taras and Bruce E. Kaufman have developed a typology of non-union employee representation arrangements in North America. See "Non-Union Employee Representation in North America: Diversity, Controversy and Uncertain Future," *Industrial Relations Journal* 37 (September 2006): 518. Joel Rodgers and Wolfgang Streeck have their own typology. See "The Study of Works Councils: Concepts and Problems," in *Works Councils: Consultation, Representation, and Cooperation in Industrial Relations*, ed. Joel Rogers and Wolfgang Streeck (Chicago: University of Chicago Press, 1995), 6–11. This variety of employee representation plans is the reason I avoid the term "company union" in this book (with the exception of its appearance in contemporary source material); when I do use it, I place quotation marks around the phrase. As David Brody has explained, "The damning term commonly used by historians, and by critics at the time, was the company union, but we will do better to accept the term advanced by employers and one more functionally descriptive—the employee representation plan (ERP), or, in some companies, works council." See Brody, "Section 8(a)(2) and the Origins of the Wagner Act," in *Labor Embattled* (Urbana: University of Illinois Press, 2005), 49.

4. Daphne Gottlieb Taras and Bruce E. Kaufman, "What Do Nonunions Do? What Should We Do about Them?" Institute for Work and Employment

Research, Task Force Working Paper #WP14, September 1, 1999, 1, http://mitsloan.mit.edu/iwer/tftaras.pdf, accessed January 22, 2007.

5. Quoted in *The New York Times*, October 1, 1915.

6. Colorado Fuel and Iron Company, "The Purposes and Principles of the Industrial Representation Plan . . . ," ca. 1920, Industrial Relations Materials, Colorado Fuel and Iron Archives, Pueblo, CO (hereafter CF&I Archives).

7. Colorado Fuel and Iron Company, "'The Present Need and the New Emphasis within Industry'. . . ," ca. 1920, ibid.

8. Quoted in *The Wilkes Barre* (PA) *Times Leader*, March 20, 1917.

9. Quoted in *The Pueblo Chieftain*, October 3, 1919.

10. Ben M. Selekman, "General Statement about Fremont County," ca. 1919, Mary Van Kleeck Papers, Box 47, Folder 6, Sophia Smith Collection, Smith College, Northampton, MA.

11. David Montgomery, *Fall of the House of Labor* (New York: Cambridge University Press, 1986), 348–350.

12. Bruce Kaufman, "Industrial Relations Counselors, Inc.: Its History and Significance," in *Industrial Relations to Human Relations and Beyond,* ed. Bruce E. Kaufman, Richard A. Beaumont, and Roy B. Helfgott (Armonk, NY: M. E. Sharpe, 2003), 55.

13. John Ensor Harr and Peter L. Johnson, *The Rockefeller Century* (New York: Charles Scribner's Sons, 1988), 62; Charles E. Harvey, "John D. Rockefeller, Jr. and the Interchurch World Movement of 1919–1920: A Different Angle on the Ecumenical Movement," *Church History* 51 (June 1982): 198–209.

14. Howard M. Gitelman, *Legacy of the Ludlow Massacre: A Chapter in American Industrial Relations* (Philadelphia: University of Pennsylvania Press, 1988), 264–265 (original emphasis).

15. Gary Dean Best, "President Wilson's Second Industrial Conference, 1919–20," *Labor History* 16 (Fall 1975): 515.

16. Jonathan Rees, *Managing the Mills: Labor Policy in the American Steel Industry during the Nonunion Era* (Lanham, MD: University Press of America, 2004), 172.

17. Samuel Gompers, "Rockefeller Organizes and Recognizes a 'Union,'" *American Federationist* 22 (November 1915): 976.

18. Bruce E. Kaufman, "Accomplishments and Shortcomings of Nonunion Employee Representation in the Pre–Wagner Act Years: A Reassessment," in *Nonunion Employee Representation: History, Contemporary Practice, and Policy*, ed. Bruce E. Kaufman and Daphne Taras (Armonk, NY: M. E. Sharpe, 2000), 25.

19. *NLRB v. Jones & Laughlin Steel Corp.*, 301 U.S. 1 (1937).

20. Bruce E. Kaufman, "The Case for the Company Union," *Labor History* 41 (August 2000): 322.

21. Of course, there are exceptions to this general rule. See, for example, Sidney Fine, *The Automobile under the Blue Eagle* (Ann Arbor: University of Michigan Press, 1963), 154–162; Ronald W. Schatz, *The Electrical Workers: A History of Labor at General Electric and Westinghouse 1923–60* (Urbana: University

of Illinois Press, 1983), 40–42, 66–67; James D. Rose, *Duquesne and the Rise of Steel Unionism* (Urbana: University of Illinois Press, 2001), 101–135.

22. Federal Council of Churches, Department of Research and Education, "Industrial Relations in the Coal Industry in Colorado," ca. 1928, 112, Josephine Roche Papers, Box 15, Folder 7, Archives, University of Colorado at Boulder Libraries, Boulder.

23. John R. Commons, *Industrial Goodwill* (New York: McGraw-Hill, 1919), 112.

24. Untitled Meeting Transcript, Canon District, November 10, 1927, 1, Industrial Relations Materials, CF&I Archives.

25. Quoted in Official Report of Proceedings of the National Labor Relations Board, "In the Matter of Colorado Fuel and Iron Corporation and Union of Mine Mill and Smelter Workers," 718, Records of the United States Court of Appeals, Tenth Circuit, Box 202, Record Group 276, National Archives and Records Administration, Rocky Mountain Region, Lakewood, CO.

CHAPTER 1: MEMORIES OF A MASSACRE

1. Bowers quoted in Priscilla Long, *Where the Sun Never Shines: A History of America's Bloody Coal Industry* (New York: Paragon House, 1989), 242.

2. Howard M. Gitelman, *Legacy of the Ludlow Massacre: A Chapter in American Industrial Relations* (Philadelphia: University of Pennsylvania Press, 1988), 242–243. See also *Trinidad Evening Picketwire*, May 31, 1918, which confirms the account Gitelman cites.

3. Quoted in *The New York Times*, April 26, 1915.

4. Rockefeller was more correct than most historians know. As Scott Martelle has written, "Rather than the intentional execution of a large number of people, the deaths seem most likely to have been the result of criminally negligent acts by the Colorado National Guard, private mine guards, and strikebreakers as they torched the camp." However, the Rockefeller Plan emerged in response to public perception of the tragedy rather than in response to what really happened, so the exact details of the massacre need not concern us here. For more, see Scott Martelle, *Blood Passion: The Ludlow Massacre and Class War in the American West* (New Brunswick, NJ: Rutgers University Press, 2007), 5 (source of the quotation in this note), 146–176.

5. Henry Demarest Lloyd, *Wealth against Commonwealth* (New York: Harper & Brothers, 1894); Ida M. Tarbell, *The History of the Standard Oil Company* (New York: McClure, Philips, 1904).

6. U.S. Congress, House, Committee on Mines and Mining, *Conditions in the Coal Mines of Colorado*, vol. 2, 63d Congress, 2nd Sess., 1914, 2874.

7. Donald McClurg, "Labor Organization in the Coal Mines of Colorado, 1878–1930." PhD dissertation, University of California–Berkeley, 1959, 320.

8. Quoted in *The New York Herald*, April 24, 1915.

9. Quoted in Martelle, *Blood Passion*, 88.

10. Coal Mine Managers, "Facts Concerning the Struggle in Colorado for Industrial Freedom," Series 1, Bulletin 1 (June 22, 1914): 19.

11. Coal Mine Managers, "Facts Concerning the Struggle in Colorado for Industrial Freedom," Series 1, Bulletin 6 (July 15, 1914): 37.

12. On Tikas, see Zeese Papanikolas, *Buried Unsung: Louis Tikas and the Ludlow Massacre* (Lincoln: University of Nebraska Press, 1991 [Salt Lake City: University of Utah Press, 1982]).

13. For details on the aftermath of the massacre, see Martelle, *Blood Passion*, 177–210; Thomas G. Andrews, *Killing for Coal: America's Deadliest Labor War* (Cambridge: Harvard University Press, 2008), 10–15.

14. *United Mine Workers Journal* 29 (May 16, 1918): 6.

15. These were just the ones for which he could find records. There may have been more.

16. Martelle, *Blood Passion*, 222–224.

17. Walter Fink, "The Ludlow Massacre," in *Massacre at Ludlow: Four Reports*, ed. Leon Stein and Philip Taft (New York: Arno and The New York Times, 1971), 19–20.

18. Martelle, *Blood Passion*, 222–224.

19. The victims in the Death Pit at Ludlow died of suffocation. One child was shot.

20. Mary Field Parton, ed., *The Autobiography of Mother Jones* (Chicago: Charles H. Kerr, 1925), 191.

21. F. Darrell Munsell, *From Redstone to Ludlow: John Cleveland Osgood's Struggle against the United Mine Workers of America* (Boulder: University Press of Colorado, 2009), 250.

22. *Rocky Mountain News* quoted in McClurg, "Labor Organization," 289–290.

23. H. Lee Scamehorn, *Mill and Mine: The CF&I in the Twentieth Century* (Lincoln: University of Nebraska Press, 1992), 18–20, 24.

24. Rockefeller quoted in Daniel Okrent, *Great Fortune: The Epic of Rockefeller Center* (New York: Penguin Books, 2003), 39.

25. Ibid.

26. John Ensor Harr and Peter J. Johnson, *The Rockefeller Century* (New York: Charles Scribner's Sons, 1988), 109–115.

27. Ivy Lee to John D. Rockefeller Jr., August 16, 1914, Rockefeller Family Archives, Record Group 2, Box 54, Folder 485, Rockefeller Archives Center (hereafter RAC), Sleepy Hollow, NY.

28. *Telluride Daily Journal*, April 22, 1914.

29. *New York Times*, April 22, 1914.

30. *Literary Digest* 69 (May 16, 1914): 1.

31. Fink, "Ludlow Massacre," 16.

32. Cited in McClurg, "Labor Organization," 289.

33. *Telluride Daily Journal*, April 21, 1914.

34. Coal Mine Managers, "The Letter to the President," May 4, 1914, Industrial Relations Materials, Colorado Fuel and Iron Archives, Pueblo, CO (hereafter CF&I Archives).

35. George S. McGovern and Leonard F. Guttridge, *The Great Coalfield War* (New York: Houghton Mifflin, 1972), 276.

36. *New York Times*, February 4, 1915.

37. Frank J. Hayes, "On Ludlow Field," *United Mine Workers Journal* 29 (June 6, 1918): 4.

38. McGovern and Guttridge, *Great Coalfield War*, 340.

39. Quoted in Barron B. Beshoar, *Out of the Depths: The Story of John R. Lawson, a Labor Leader* (Denver: Golden Bell, 1942), 316. Beshoar is the source of all details about the Lawson case unless otherwise noted.

40. John Thomas Hogle, "The Rockefeller Plan: Workers, Managers and the Struggle over Unionism in Colorado Fuel and Iron, 1915–1942." PhD dissertation, University of Colorado–Boulder, 1992, 101.

41. Frank Walsh, "Mr. Walsh at People's Power League, May 2, 1915," Office of the Messrs. Rockefeller (OMR) Economic Interests, Record Group III 2 F, Box 23, Folder 206, RAC.

42. *The New York Times*, June 21, 1915.

43. John D. Rockefeller Jr., "Statements of John D. Rockefeller Jr. before the United States Commission on Industrial Relations," May 20–21, 1915, 8, OMR Economic Interests, Record Group II 2 F, Box 23, Folder 217, RAC.

44. *United Mine Workers Journal* 29 (May 16, 1918): 6.

45. United States Department of Labor, "Proceedings of the First Industrial Conference," October 6–23, 1919 (Washington, DC: Government Printing Office, 1920), 144.

46. See Beshoar, *Out of the Depths*, 209–210; *The New York Times*, August 19, 1914.

47. Even after the furor over the Ludlow Massacre had died down, Upton Sinclair continued to attack John D. Rockefeller Jr. In 1917 he published a novel entitled *King Coal*. In a postscript he wrote about his research in the third person: "Most of the details of this picture were gathered in [Colorado], which the writer visited on three occasions during and just after the great coal-strike of 1913–14. The book gives a true picture of conditions and events observed by him at that time. Practically all the characters are real persons, and every incident which has social significance is not merely a true incident, but a typical one." The Rockefeller character is named Percy Harrigan. He gets to utter some memorable lines: "'The world can't stop moving just because there's been a mine-disaster,' said the Coal King's son. 'People have engagements they must keep.'" See Upton Sinclair, *King Coal: A Novel* (New York: Macmillan, 1917), 384, 265. Sinclair wrote another book about the 1913–1914 Colorado strike called *The Coal War*. Elliot Gorn has called it "a dramatized history of the events of 1914" that fulfilled Sinclair's wish to "tell the specific story of the Colorado strike." Gorn, *Mother Jones: The Most Dangerous Woman in America* (New York: Hill and Wang, 2001),

370, n. 1. Sinclair did not find a publisher for the book, so it was not released until 1976, long after memories of the strike had faded.

48. Ivy Lee to John D. Rockefeller Jr., August 16, 1914, RAC.

49. United Mine Workers of America, District 15 Policy Committee, "The Struggle in Colorado for Industrial Freedom," Bulletin 6, September 16, 1914, 2.

50. John Fitch, "What Rockefeller Knew and What He Did," *The Survey* (August 21, 1915): 468.

51. Gitelman, *Legacy of the Ludlow Massacre*, 59. For more on the controversy surrounding Lee, see Kirk Hallahan, "Ivy Lee and the Rockefellers' Response to the 1913–1914 Colorado Coal Strike," *Journal of Public Relations Research* 14 (2002): 301–303.

52. Coal Mine Managers, "Facts Concerning the Struggle in Colorado for Industrial Freedom," Series 1, Bulletin 8, July 25, 1914, n.p.

53. "Testimony of John D. Rockefeller Jr. . . ," 267, RG 2.1, CF&I Archives.

54. George P. West, "Report on the Colorado Strike," in *Massacre at Ludlow: Four Reports*, ed. Leon Stein and Philip Taft (New York: Arno and The New York Times, 1971), 155.

55. Cited in Ben M. Selekman and Mary Van Kleeck, *Employes' Representation in Coal Mines* (New York: Russell Sage Foundation, 1924), 14–15.

56. Raymond B. Fosdick, "RBF Conversations with JDR Jr.," 37, Rockefeller Family Archives, JDR Jr. Personal, Box 57, Folder 503, RAC.

57. Ivy Lee to John D. Rockefeller Jr., August 16, 1914, RAC.

58. Selekman and Van Kleeck, *Employes' Representation*, 12.

59. Jesse Welborn to Woodrow Wilson, September 18, 1914, Jesse Floyd Welborn Papers, Box 1, Folder 2, Colorado Historical Society, Denver (hereafter Welborn Papers).

60. Cited in Beshoar, *Out of the Depths*, 245.

61. Murphy excerpted in West, "Report on the Colorado Strike," 174.

62. Cited in Beshoar, *Out of the Depths*, 246.

63. Mike Livoda Interview, August 1975, Eric Margolis Coal Project (Oral History Interviews), 7, Archives, University of Colorado at Boulder Libraries, Boulder.

64. Jesse Welborn to Seth Low et al., February 1, 1915, Box 1, Folder 9, Welborn Papers. No record seems to exist of this Denver meeting ever occurring.

65. Hogle, "Rockefeller Plan," 89–90.

66. King quoted in Fred A. McGregor, *The Fall and Rise of Mackenzie King* (Toronto: Macmillan, 1962), 159.

67. Quoted in *The Denver Times*, September 28, 1915.

68. Fitch, "What Rockefeller Knew," 462.

69. Rockefeller quoted in McGovern and Guttridge, *Great Coalfield War*, 319.

70. Quoted in *The New York Herald*, January 26, 1915.

71. Quoted in *The New York Times*, January 28, 1915.

NOTES

72. Jones quoted in Gorn, *Mother Jones*, 224.

73. Rockefeller quoted in Ron Chernow, *Titan: The Life of John D. Rockefeller, Sr.* (New York: Random House, 1998), 588.

74. John D. Rockefeller Jr. to John D. Rockefeller Sr., September 15, 1915, in *"Dear Father"/"Dear Son": Correspondence of John D. Rockefeller and John D. Rockefeller, Jr.*, ed. Joseph W. Ernst (New York: Fordham University Press, 1994), 58.

75. McGregor, *Fall and Rise of Mackenzie King*, 176–177.

76. Munsell, *From Redstone to Ludlow*, 327.

77. Rockefeller Jr. to Rockefeller Sr., 61–62. Rockefeller was not being more flexible in public while plotting in private. "Please destroy this letter," he wrote at the end of the text, since this private position was actually much more liberal than any he had taken in public to that point.

78. *New York Times*, October 3, 1915.

79. Quoted in *The Denver Post*, September 24, 1915.

80. *Denver Times*, September 24, 1915.

81. Louise Reynolds, *Mackenzie King: Friends & Lovers* (Victoria, BC: Trafford, 2005), 148.

82. Quoted in *The Denver Post*, September 20, 1915.

83. John D. Rockefeller Jr., "Address by John D. Rockefeller Jr.," May 15, 1916, 4, Rockefeller Family Archives, Record Group III 2 Z, Box 1, Folder 49, RAC.

84. Fosdick, "RBF Conversations with JDR Jr.," 47A.

85. *Denver Post*, September 21, 1915.

86. *Chicago Daily Tribune*, September 26, 1915.

87. Quoted in *The Denver Times*, September 23, 1915.

88. Ibid., September 22, 1915.

89. *New York American*, September 24, 1915.

90. Quoted in Eric Margolis, "Western Coal Mining as a Way of Life," *Journal of the West* 24 (July 1985): 104.

91. *Denver Post*, October 3, 1915.

92. *Fort Collins Weekly Courier*, October 3, 1915.

93. John D. Rockefeller Jr., "To the People of Colorado," in "The Colorado Industrial Plan" (privately printed, 1916), 60, 62.

94. *New York Evening Post*, November 6, 1915.

95. Selekman and Van Kleeck, *Employes' Representation*, 27.

96. Ibid., 80–81.

97. Quoted in *The Akron* (Ohio) *Weekly Pioneer Press*, October 8, 1915.

98. See John D. Rockefeller Jr., Memorandum Concerning Colorado Trip, 1915, 33, Rockefeller Family Archives, Record Group III 2 C, Box 22, Folder 198, RAC.

CHAPTER 2: STUDENT AND TEACHER

1. John D. Rockefeller Jr. to Harry Emerson Fosdick, quoted in Peter Collier and David Horowitz, *The Rockefellers* (New York: Holt Rinehart Winston, 1976), 153–154.

2. John D. Rockefeller Jr., "Address of John D. Rockefeller Jr. . . . ," *Colorado Fuel and Iron Company Industrial Bulletin* 1 (October 1915): 11–12 (emphasis added).

3. Ibid., October 2 and 8, 1915, 18–19.

4. David Fairris, "From Exit to Voice in Shopfloor Governance: The Case of Company Unions," *Business History Review* 69 (Winter 1995): 494–529.

5. Richard B. Freeman and Edward P. Lazear, "An Economic Analysis of Works Councils," in *Works Councils*, An Economic Analysis of Works Councils (1994-11-01). National Bureau of Economic Research Working Paper W4918. Available at SSRN: http://ssrn.com/abstract=227953.

6. Richard B. Freeman, "The Exit-Voice Tradeoff in the Labor Market: Unionism, Job Tenure, Quits and Separations," *Quarterly Journal of Economics* 94 (June 1980): 647.

7. David Rockefeller, *Memoirs* (New York: Random House, 2002), 21.

8. Raymond Fosdick, "The Mackenzie King Correspondence," Research Notes, ca. 1956, Rockefeller Family Archives, Record Group III 2 Z, Box 53, Folder 481, Rockefeller Archives Center (hereafter RAC), Sleepy Hollow, NY. Fosdick's research notes include copies of Rockefeller's and Rockefeller-related correspondence copied out in full. When I could not find a letter in another source, I cite the location of the copies in this record group from the RAC.

9. As Rockefeller wrote to King in 1915, "I am perfectly sincere when I say that had it not been for your wise guidance and counsel, I should not only have made mistakes, but would have failed to have accomplished much of what you have made possible. It has been a great joy to work shoulder to shoulder with you and Mr. Welborn in such perfect harmony and with such mutual understanding and confidence, as we have all sought to make what contribution we could toward the advancement of industrial peace and the promotion of good will and a better understanding on the part of the people of Colorado." See John D. Rockefeller Jr. to Mackenzie King, October 10, 1915, JDR Jr. Personal, ibid.

10. Mark Twain, "Mr. Rockefeller's Bible Class," March 20, 1906, in *Mark Twain in Eruption*, ed. Bernard DeVoto (New York: Harper & Brothers, 1922), 84.

11. Ben M. Selekman and Mary Van Kleeck, *Employes' Representation in Coal Mines* (New York: Russell Sage Foundation, 1924), 6.

12. John D. Rockefeller Jr. to Fred Arnold, March 7, 1919, Rockefeller Family Archives, Record Group II 2 F, Box 20, Folder 170, RAC.

13. *The Denver Times*, September 28, 1915.

14. Raymond B. Fosdick, *John D. Rockefeller Jr.: A Portrait* (New York: Harper & Brothers, 1956), 1.

15. David Rockefeller, *Memoirs*, 19.

16. Mackenzie King to Abby Aldrich Rockefeller, 1915, quoted in Bernice Kert, *Abby Aldrich Rockefeller: The Woman in the Family* (New York: Random House, 1993), 151.

17. Bruce E. Kaufman, *The Global Evolution of Industrial Relations: Events, Ideas and the IIRA* (Geneva: International Labour Office, 2004), 133.

18. Rockefeller quoted in Ron Chernow, *Titan: The Life of John D. Rockefeller, Sr.* (New York: Random House, 1998), 572.

19. Howard M. Gitelman, *Legacy of the Ludlow Massacre: A Chapter in American Industrial Relations* (Philadelphia: University of Pennsylvania Press, 1987), 15.

20. Charles M. Schwab quoted in Kenneth Warren, *Industrial Genius: The Working Life of Charles Michael Schwab* (Pittsburgh: University of Pittsburgh Press, 2007), 191, 190.

21. Gitelman, *Legacy of the Ludlow Massacre*, 254.

22. U.S. House of Representatives, "Report of the Colorado Strike Investigation" (Washington, DC: Government Printing Office, 1915), 39; excerpted in *Massacre at Ludlow: Four Reports*, ed. Leon Stein and Philip Taft (North Stratford, NH: Ayer, 2000), 39.

23. Mina Pendo to Raymond Fosdick, October 29, 1954, Rockefeller Family Archives, Record Group III 2 Z, Box 53, Folder 480, RAC.

24. "Testimony of John D. Rockefeller Jr. Given before the United States Commission on Industrial Relations," 1915, 67–68, Colorado Fuel and Iron Archives, Pueblo, CO (hereafter CF&I Archives).

25. Quoted in *The Lexington* (Kentucky) *Herald*, November 21, 1915.

26. Charles A. Heydt to E. H. Weitzel, December 18, 1915, Rockefeller Family Archives, Business Interests, Box 22, Folder 198, RAC.

27. J. F. Welborn to John D. Rockefeller Jr., December 7, 1917, Industrial Relations Materials, CF&I Archives. Selekman and Van Kleeck suggest that Rockefeller was still involved in settling grievances under the plan as late as 1919. See Selekman and Van Kleeck, *Employes' Representation*, 174, n. 1.

28. John D. Rockefeller Jr. to Clarence Hicks, December 22, 1915, Industrial Relations Materials, CF&I Archives.

29. John D. Rockefeller Jr. to Clarence Hicks, July 18, 1916, ibid.

30. John D. Rockefeller Jr. to John D. Rockefeller Sr., September 15, 1915, in *"Dear Father"/"Dear Son": Correspondence of John D. Rockefeller and John D. Rockefeller, Jr.*, ed. Joseph W. Ernst (New York: Fordham University Press, 1994), 61.

31. John D. Rockefeller Jr. to A. H. Lichty, April 13, 1921, Rockefeller Family Archives, Record Group II 2 F, Box 15, Folder 124, RAC.

32. Quoted in *The Denver Post*, June 6, 1918.

33. Quoted in *The New York Tribune*, January 27, 1915.

34. Rockefeller quoted in Daniel Okrent, *Great Fortune: The Epic of Rockefeller Center* (New York: Penguin, 2003), 390.

35. Rockefeller quoted in Chernow, *Titan*, 582.
36. Ibid.
37. Mackenzie King Diary, December 31, 1914, 304, Collections Canada, http://king.collectionscanada.ca.
38. H. S. Ferns and B. Ostry, *The Age of Mackenzie King: The Rise of the Leader* (London: William Heinemann, 1955), 212.
39. F. Darrell Munsell, *From Redstone to Ludlow: John Cleveland Osgood's Struggle against the United Mine Workers of America* (Boulder: University Press of Colorado, 2009), 299.
40. Mackenzie King Diary, December 31, 1914, 17.
41. In fact, in appreciation for King's assistance during his years out of government, Rockefeller gave him a gift of $100,000 in stock after he retired. See Ferns and Ostry, *Age of Mackenzie King*, 215.
42. Ben M. Selekman, *Postponing Strikes: A Study of the Industrial Disputes Investigations Act of Canada* (New York: Russell Sage Foundation, 1927), 48. The Canadian Supreme Court found the law unconstitutional in 1925.
43. King quoted in Stephen J. Scheinberg, *Employers and Reformers: The Development of Corporation Labor Policy, 1900–1940* (New York: Garland, 1986), 151.
44. Gitelman, *Legacy of the Ludlow Massacre*, 61.
45. Raymond B. Fosdick, "RBF Conversations with JDR Jr.," 76, Rockefeller Family Archives, JDR Jr. Personal, Record Group III 22, Box 57, Folder 503, RAC.
46. John D. Rockefeller Jr. to J. F. Welborn, November 7, 1919, Rockefeller Family Archives, Record Group 2, Box 54, Folder 486, RAC.
47. Mackenzie King to John D. Rockefeller Jr., January 17, 1921, Rockefeller Family Archives, Record Group III 2 C, Box 16, Folder 134, RAC.
48. Ruth O'Brien, *Labor's Paradox: The Republican Origins of New Deal Labor Policy, 1886–1935* (Chapel Hill: University of North Carolina Press, 1998), 119.
49. Fosdick, "RBF Conversations with JDR Jr.," 87.
50. Bowers excerpted in George P. West, "Report on the Colorado Strike," U.S. Industrial Relations Commission, Washington, DC, 1915, 169; excerpted in *Massacre at Ludlow: Four Reports*, ed. Leon Stein and Philip Taft (North Stratford, NH: Ayer, 2000), 169.
51. John Ensor Harr and Peter J. Johnson, *The Rockefeller Century* (New York: Charles Scribner's Sons, 1988), 139–140.
52. J. F. Welborn to Mackenzie King, February 2, 1916, Jesse Floyd Welborn Papers, Box 1, Folder 10, Colorado Historical Society, Denver.
53. Fosdick, "RBF Conversations with JDR Jr.," 88.
54. Fred A. McGregor, *The Fall and Rise of Mackenzie King* (Toronto: Macmillan, 1962), 159.
55. J. F. Welborn to Mackenzie King, December 20, 1919, Rockefeller Family Archives, Record Group III 2 C, Box 15, Folder 127A, RAC.
56. Federal Council of Churches, Department of Research and Education, "Industrial Relations in the Coal Industry in Colorado," ca. 1928, 115, Josephine

Roche Papers, Box 15, Folder 4, Archives, University of Colorado at Boulder Libraries, Boulder.

57. Mackenzie King Testimony, National Industrial Conference (Canada), *Official Report of the Proceedings and Discussions* (Ottawa: September 15–20, 1919), 160.

58. Gitelman, *Legacy of the Ludlow Massacre*, 167.

59. Mackenzie King, Untitled Memo, ca. 1919, Rockefeller Family Archives, Record Group 2 F, Box 15, Folder 121, RAC.

60. John D. Rockefeller Jr., "Statement of John D. Rockefeller Jr. before the United States Commission on Industrial Relations," January 25, 1915, 3–4, Rockefeller Family Archives, Record Group II 2 F, Box 23, Folder 210, RAC.

61. John D. Rockefeller Jr. to E. H. Weitzel, March 13, 1916, Rockefeller Family Archives, Record Group II 2 F, Box 15, Folder 127, RAC.

62. Quoted in *The Denver Times*, October 1, 1915. Speaking to biographer Raymond Fosdick in 1953, Rockefeller offered another reason: "We felt that having intimate relations with the immediate employees was better than through a national union. I still think so, and I became a champion of the idea—an advocate of some plan by which working men could get through the barriers of local bosses to air their grievances to the men on top." In other words, he felt CF&I employees had made the wrong choice. See Fosdick, "RBF Conversations with JDR Jr.," 47A.

63. Melvyn Dubofsky and Warren Van Tine, *John L. Lewis: A Biography* (New York: Quadrangle/The New York Times, 1977), xv.

64. John D. Rockefeller Jr., "Remarks of John D. Rockefeller Jr. at the National Industrial Conference," pamphlet, October 16, 1919, Washington, DC, 11, RAC.

65. John D. Rockefeller Jr. to Robert H. Adams, May 13, 1919, Rockefeller Family Archives, Record Group III 2 Z, Box 53, Folder 480, RAC.

66. King quoted in Fosdick, *John D. Rockefeller Jr.*, 170.

67. Mackenzie King to Starr J. Murphy, December 11, 1919, Rockefeller Family Archives, Record Group II 2 F, Box 14, Folder 113, RAC.

68. Mackenzie King Diary, September 29, 1919, 272.

69. "Extract from a Letter from Starr J. Murphy," April 23, 1919, Rockefeller Family Archives, Record Group II 2 F, Box 14, Folder 108, RAC.

70. Nelson Lichtenstein and Howell John Harris, "Introduction: A Century of Industrial Democracy in America," in *Industrial Democracy: The Ambiguous Promise*, ed. Nelson Lichtenstein and Howell John Harris (New York: Cambridge University Press, 1993), 9.

71. Fosdick, *John D. Rockefeller Jr.*, 167.

72. "Memorandum of Distribution of 500,000 Copies of the Booklet Entitled 'The Colorado Industrial Plan,'" Rockefeller Family Archives, Record Group III 2 C, Box 11, Folder 93, RAC.

73. Ibid., 18.

74. John D. Rockefeller Jr., "Labor and Capital—Partners," *Atlantic Monthly* (January 1916): 9, 5, 28.

75. *Pueblo Chieftain*, June 5, 1918.

76. John D. Rockefeller Jr. to John D. Rockefeller Sr., June 6, 1918, "Dear Father"/"Dear Son," 83.

77. Quoted in *The New York Sun*, August 12, 1918.

78. John D. Rockefeller Jr., "Patriotism and Industry," June 4, 1918, Rockefeller Family Archives, Record Group III 2 Z, Box 2, Folder 68, RAC.

79. John D. Rockefeller Jr., "Brotherhood of Men and Nations," pamphlet, June 13, 1918, 11, RAC.

80. John D. Rockefeller Jr., "Representation in Industry," Address before the War Emergency and Reconstruction Conference, Atlantic City, NJ, December 5, 1918, 6, RAC.

81. U.S. Department of Labor, "Proceedings of the First Industrial Conference," October 6–23, 1919 (Washington, DC: Government Printing Office, 1920), 143.

82. On events at President Wilson's Industrial Conference, see Gitelman, *Legacy of the Ludlow Massacre*, 315–318.

83. Gilbert L. Lacher, "Industrial Relations Plan Stands Time Test," *Iron Age* 115 (January 29, 1925): 6.

84. H. Lee Scamehorn, *Mill & Mine: The CF&I in the Twentieth Century* (Lincoln: University of Nebraska Press, 1992), 126–132.

85. Curtis, Fosdick, and Belknap, Report on Industrial Relations in the Colorado Fuel and Iron Company, ca. 1924, 21, Industrial Relations Materials, CF&I Archives.

CHAPTER 3: BETWEEN TWO EXTREMES

1. Mackenzie King to John D. Rockefeller Jr., August 6, 1914, excerpted in George P. West, "Report on the Colorado Strike," in *Massacre at Ludlow: Four Reports*, ed. Leon Stein and Philip Taft (New York: Arno and The New York Times, 1971), 163. I have seen originals of some of these letters in the Welborn Papers at the Colorado Historical Society and the Records of the Rockefeller Business Interests at the Rockefeller Archives Center. Because the transcripts in West are easier to access, I cite that source when possible.

2. Mary Field Parton, ed., *The Autobiography of Mother Jones* (Chicago: Charles H. Kerr, 1925), 201.

3. Wagner quoted in Philip Taft, *The A.F. of L. from the Death of Gompers to the Merger* (New York: Harper & Brothers, 1959), 122.

4. Daniel Nelson, "The Company Union Movement, 1900–1937: A Reexamination," *Business History Review* 56, 3 (Autumn 1982): 341.

5. F. Darrell Munsell, *From Redstone to Ludlow: John Cleveland Osgood's Struggle against the United Mine Workers of America* (Boulder: University Press of Colorado, 2009), 351.

6. William L. Mackenzie King to John D. Rockefeller Jr., August 6, 1914, excerpted in West, "Report on the Colorado Strike," 163.

7. Daniel Nelson, "The AFL and Company Unionism," in *Nonunion Employee Representation: History, Contemporary Practice and Policy*, ed. Bruce E. Kaufman and Daphne Gottlieb Taras (Armonk, NY: M. E. Sharpe, 2000), 72.

8. *Leslie's Illustrated Weekly Newspaper* 121 (October 21, 1915): 437.

9. American Federation of Labor, *Report of Proceedings of the Thirty-Ninth Annual Convention of the American Federation of Labor* (Washington, DC: Law Reporter Printing, 1919), 303.

10. Larry G. Gerber, "Corporatism in Comparative Perspective: The Impact of the First World War on American and British Labor Relations," *Business History Review* 62 (Spring 1988): 100–101.

11. I am borrowing an argument from Colin Gordon, *New Deals: Business, Labor and Politics in America, 1920–1935* (New York: Cambridge University Press, 1994), 123.

12. The original version of the plan for the miners is the version I cite unless otherwise noted. The entire text of the plan is Appendix 1. I will not footnote passages from this version of the plan, as they can all be found there. Later versions, including the consolidated version of the plan, are available in Ben M. Selekman and Mary Van Kleeck, *Employes' Representation in Coal Mines* (New York: Russell Sage Foundation, 1924). For the first version of the plan in the steel mill, see the appendixes in Ben M. Selekman, *Employes' Representation in Steel Works* (New York: Russell Sage Foundation, 1924). As we will see in Chapter 8, the plan was revised again in 1937 and 1938. The nature of those revisions is discussed in that chapter.

13. Selekman and Van Kleeck, *Employes' Representation*, 145, n. 1.

14. Management added tenth and eleventh divisions in 1919. Selekman, *Employes' Representation in Steel Works*, 237. Not even the consolidated version of the plan addressed the question of whether representatives from the mines and the mill could meet with management at the same time.

15. Howard M. Gitelman, *Legacy of the Ludlow Massacre: A Chapter in American Industrial Relations* (Philadelphia: University of Pennsylvania Press, 1988), 192.

16. U.S. Department of Labor, "Characteristics of Company Unions 1935," Bulletin 634, June 1937, 140.

17. American Federation of Labor, *Report of Proceedings*, 302.

18. If an employee representative resigned, a special election would be held in that division to fill the spot.

19. B. J. Matteson to William J. Aitken, May 10, 1924, Industrial Relations Materials, Colorado Fuel and Iron Archives, Pueblo, CO (hereafter CF&I Archives).

20. Welborn excerpted in West, "Report on the Colorado Strike," 182.

21. Mackenzie King Diary, December 31, 1914, 562, Collections Canada, http://king.collectionscanada.ca.

NOTES

22. Gilbert L. Lacher, "Industrial Relations Plan Stands Test of Time," *Iron Age* 115 (January 29, 1925): 6.

23. American Federation of Labor, *Report of Proceedings,* 302.

24. Colorado Fuel and Iron Company, "The Purposes and Principles of the Industrial Representation Plan . . . ," ca. 1920, Industrial Relations Materials, CF&I Archives.

25. Testimony of Warren Densmore, Official Report of Proceedings before the National Labor Relations Board, Case nos. XXII-C-127–134 and XXII-R-33, vol. 5 (July 9, 1938): 569, National Labor Relations Board, Administrative Files and Dockets 1935–, Record Group 25, National Archives II, College Park, MD.

26. Ben M. Selekman, Interview with John Merritt, December 8, 1919, Mary Van Kleeck Papers, Box 47, Folder 6, Sophia Smith Collection, Smith College, Northampton, MA.

27. Colston E. Warne and Merrill E. Gaddis, "Eleven Years of Compulsory Investigation of Industrial Disputes in Colorado," *Journal of Political Economy* 35 (October 1927): 657.

28. Colorado Industrial Commission, *First Report of the Industrial Commission of Colorado 1915–1917*, 87, Department of Labor and Employment, Industrial Commission, Box 28617, Colorado State Archives, Denver.

29. Part III, Section 16 of the consolidated plan stated that if the Committee on Industrial Cooperation, Conciliation and Wages could not agree on a solution to a dispute, the vote of a majority of members of that committee could send the issue to an impartial umpire. Perhaps because the composition of the committee was split evenly between labor and management, no available evidence suggests that this situation ever arose.

30. David Brody, "Section 8a(2) and the Origins of the Wagner Act," in *Labor Embattled: History, Power, Rights* (Urbana: University of Illinois Press, 2005), 51.

31. Quoted in David Brody, *Steelworkers in America: The Nonunion Era* (Urbana: University of Illinois Press, 1998 [Cambridge: Harvard University Press, 1960]).

32. Lacher, "Industrial Relations Plan Stands Test of Time," 4. Even when workers won these workmen's compensation disputes, they could not have been happy that management had made them hire a lawyer and wait for their payout.

33. "History of Companies Predecessor to the Colorado Fuel and Iron Company," ca. 1925, 126, CF&I Archives.

34. See, for example, Brody, *Steelworkers in America*, 81.

35. Mackenzie King Diary, December 31, 1914, 694.

36. John D. Rockefeller Jr. to J. F. Welborn, February 10, 1916, Rockefeller Family Archives, Record Group III 2 Z, Box 54, Folder 482, Rockefeller Archives Center (hereafter RAC), Sleepy Hollow, NY. The manner in which the company adhered to Rockefeller's principled stand with respect to the mines is traced in Chapter 5.

NOTES

37. Quoted in Selekman and Van Kleeck, *Employes' Representation*, 431.

38. Selekman, *Employes' Representation in Steel Works*, 122.

39. The terms of the Memoranda of Agreement for both miners and steelworkers carried over to the consolidated version of the plan upon its passage in 1921.

40. Selekman and Van Kleeck, *Employes' Representation*, 236–241. When the steelworkers first had the opportunity to ratify the plan in 1916, they received a memorandum of their own to sign. Its provisions with respect to living conditions were much like those in the agreement for the coal miners. It covered rent, coal, bathhouses, and the like. But its provision on working conditions created a more direct relationship with the competition. Instead of limiting what would be determined in relation to CF&I's competitors to wages (as the agreement with the miners did), this memorandum specifically linked working conditions in Colorado with those in the East.

41. See, for example, Stuart Brandes, *American Welfare Capitalism* (Chicago: University of Chicago Press, 1976).

42. Frank J. Weed, "The Sociological Department at the Colorado Fuel and Iron Company, 1901 to 1907: Scientific Paternalism and Industrial Control," *Journal of the History of the Behavioral Sciences* 41 (Summer 2005): 274–276.

43. Ibid., 277.

44. H. Lee Scamehorn, *Mine & Mill: The CF&I in the Twentieth Century* (Lincoln: University of Nebraska Press, 1992), 85.

45. J. F. Welborn, "To All Concerned," October 23, 1915, Colorado Fuel and Iron Collection, Box 12, Colorado Historical Society, Denver.

46. Colorado Fuel and Iron Company, "Twenty-Fourth Annual Report," September 1916, 5, CF&I Archives.

47. Mackenzie King Diary, March 7–9, 1917, 12649.

48. H. B. Carpenter, "Your Institution," *The Steel Works Blast*, March 21, 1927.

49. Thomas Winter, *Making Men, Making Class: The YMCA and Workingmen, 1877–1920* (Chicago: University of Chicago Press, 2002), 41.

50. John D. Rockefeller Jr. to John D. Rockefeller Sr., January 18, 1917, in *"Dear Father"/"Dear Son,"* ed. Joseph W. Ernst (New York: Fordham University Press, 1994), 71.

51. Quoted in *Colorado Fuel and Iron Company Industrial Bulletin*, December 22, 1915.

52. YMCA Industrial Department, "Among the Coal Miners," 1917, n.p., Kautz YMCA Archives, University of Minnesota, Minneapolis.

53. Charles R. Towson to John D. Rockefeller Jr., May 23, 1919, Rockefeller Family Archives, Record Group III 2 C, Box 18, Folder 156, RAC.

54. A. H. Lichty to John D. Rockefeller Jr., March 22, 1922, ibid.

55. On welfare capitalism at U.S. Steel, see Jonathan Rees, *Managing the Mills: Labor Policy in the American Steel Industry during the Nonunion Era* (Lanham, MD: University Press of America, 2003), 101–141.

NOTES

56. R. W. Corwin, "Advances in Medical Department Service," *Colorado Fuel and Iron Company Industrial Bulletin* 4 (January 15, 1919): 7.

57. The copies of *Industrial Bulletin*s in the Industrial Department Records at the Kautz YMCA Archives at the University of Minnesota in Minneapolis include a card for disinterested "friends" of the company to take themselves off the mailing list. The company was worried about higher printing costs. It did not sell the magazine by subscription.

58. The evidence for this point is discussed at length in Chapter 8.

59. Denise Pan, "Peace and Conflict in an Industrial Family: Company Identity and Class Consciousness in a Multi-ethnic Community, Colorado Fuel and Iron's Cameron and Walsen Camps, 1913–1928," MA thesis, University of Colorado–Boulder, 1994, 86.

60. Cited in "District Conference of Steel Works Representatives Held at Pueblo, Colorado," May 6, 1921, Industrial Relations Materials, CF&I Archives.

61. Curtis, Fosdick, and Belknap, Report on Industrial Relations at the Colorado Fuel and Iron Company, ca. 1924, 84, ibid.

62. "August Committee Meetings," *Colorado Fuel and Iron Company Industrial Bulletin* 3 (September 7, 1918): n.p.

63. Selekman and Van Kleeck, *Employes' Representation,* 145, n. 1.

64. A. J. Diamond, "The Industrial Representation Plan: An Appreciation," *Colorado Fuel and Iron Company Industrial Bulletin* 4 (January 15, 1919): 14.

65. Cited in Minutes of the Joint Conference of Employee Representatives for the Minnequa Works, February 15, 1918, Industrial Relations Materials, CF&I Archives.

66. Quoted in Second Joint Conference of Minnequa Steel Works Representatives . . . , May 28, 1920, ibid.

67. Quoted in Third Joint Conference of Minnequa Steel Works Representatives . . . , September 29, 1920, ibid.

68. Rockefeller not only bankrolled his trip to Colorado but also donated money to Harvard to pay Mayo's salary. See Kyle Bruce and Chris Nyland, "Democracy or Seduction? The Demonization of Scientific Management and the Deification of Human Relations," 14. Paper presented at the conference "The Right and Labor," Santa Barbara, CA, January 2009.

69. Elton Mayo to Colonel Arthur Woods, November 20, 1928, 9–10, Elton Mayo Papers, HBS Archives, Box 3b, Baker Library Historical Collections, Harvard Business School, Boston, MA. Despite Mayo's advice, CF&I continued to print the supplementary bulletins that contained minutes until at least 1938.

70. Ibid.

71. Curtis, Fosdick, and Belknap, Report on Industrial Relations: Summary and Conclusions, n.p., CF&I Archives.

72. Greg Patmore, "Employee Representation Plans in North America and Australia, 1915–1935: An Employer Response to Workplace Democracy," *WorkSite: Issues in Workplace Relations,* http://www.econ.usyd.edu.au/wos/worksite/employer.html, accessed October 30, 2006.

73. Joseph A. McCartin, *Labor's Great War: The Struggle for Industrial Democracy and the Origins of Modern Labor Relations, 1912–1921* (Chapel Hill: University of North Carolina Press, 1997), 88.

74. Milton Derber, *The American Idea of Industrial Democracy, 1865–1965* (Urbana: University of Illinois Press, 1970), 213.

75. McCartin, *Labor's Great War*, 100.

76. U.S. Department of Labor, National War Labor Board, *Report of the Secretary of the National War Labor Board to the Secretary of Labor for the Twelve Months Ending May 31, 1919* (Washington, DC: Government Printing Office, 1920), 56.

77. National War Labor Board quoted in Rees, *Managing the Mills*, 163.

78. Stephen J. Scheinberg, *Employers and Reformers: The Development of Corporation Labor Policy, 1900–1940* (New York: Garland, 1986), 165.

79. Gary Dean Best, "President Wilson's Second Industrial Conference, 1919–20," *Labor History* 16 (Fall 1975): 514.

80. Sumner Slichter quoted in Bruce E. Kaufman, "Accomplishments and Shortcomings of Nonunion Employee Representation in the Pre–Wagner Act Years: A Reassessment," in *Nonunion Employee Representation: History, Contemporary Practice, and Policy*, ed. Bruce E. Kaufman and Daphne Gottlieb Taras (Armonk, NY: M. E. Sharpe, 2000), 45.

81. Grant N. Farr, "The Origins of Recent Labor Policy," University of Colorado Press Series in Economics 3 (Boulder: University of Colorado Press, 1959), 17.

82. Sherman Rogers, "Industrial Representation: Success or Failure?" *The Outlook* 128 (August 31, 1921): 691.

83. Sanford Jacoby, *Employing Bureaucracy: Managers, Unions, and the Transformation of Work in American Industry, 1900–1945* (New York: Columbia University Press, 1985), 187–188.

84. Richard B. Freeman, "The Exit-Voice Tradeoff in the Labor Market: Unionism, Job Tenure, Quits and Separation," *Quarterly Journal of Economics* 94 (June 1980): 647.

CHAPTER 4: DIVISIONS IN THE RANKS

1. Ben M. Selekman, *Employes' Representation in Steel Works* (New York: Russell Sage Foundation, 1924), 172.

2. T. J. Brown, "Andrew Diamond," *The Steel Works Blast*, July 23, 1923, 1; Official Report of the Proceedings before the National Labor Relations Board, "In the Matter of the Colorado Fuel and Iron Corporation and International Union of Mine, Mill and Smelter Workers," 436–437, Records of the United States Courts of Appeals, Tenth Circuit, Box 202, Record Group 276, National Archives and Records Administration, Rocky Mountain Region, Lakewood, CO.

3. Telephone interview with Rose Ledbetter, November 12, 2007.

NOTES

4. A. J. Diamond, "The Industrial Representation Plan: An Appreciation," *Colorado Fuel and Iron Company Industrial Bulletin* 4 (January 15, 1919): 14.

5. Andrew Diamond to B. J. Matteson, January 14, 1924, Industrial Relations Materials, Colorado Fuel and Iron Archives, Pueblo, CO (hereafter CF&I Archives).

6. Elton Mayo to Colonel Arthur Woods, November 20, 1928, 30–31, Elton Mayo Papers, HBS Archives, Box 3b, Folder 18, Baker Library Historical Collections, Harvard Business School, Boston, MA (hereafter Mayo Papers).

7. Ibid., 19.

8. Eric Margolis, "Western Coal Mining as a Way of Life," *Journal of the West* 24 (July 1985): 104–105.

9. Powers Hapgood, "Paternalism versus Unionism in Mining Camps," *The Nation* 112 (May 4, 1921): 662.

10. Federal Council of Churches, Department of Research and Education, "Industrial Relations in the Coal Industry in Colorado," 122, Josephine Roche Papers, Box 15, Folder 7, Archives, University of Colorado at Boulder Libraries, Boulder.

11. Richard White, *"It's Your Misfortune and None of My Own": A New History of the American West* (Norman: University of Oklahoma Press, 1991), 296.

12. Gilbert L. Lacher, "Industrial Relations Plan Stands Test of Time," *Iron Age* 115 (January 29, 1925): 6.

13. Although I recognize it as an anachronism, I use the term "Mexican American" to refer to second- or third-generation workers of Spanish or Mexican backgrounds. When I use the term "Mexican," it will refer to recent Mexican immigrants or migrants. However, if the term "Mexican" appears in a quoted source, it might refer to people from either of these groups.

14. Sarah Deutsch, *No Separate Refuge: Culture, Class and Gender on an Anglo-Hispanic Frontier in the American Southwest, 1880–1940* (New York: Oxford University Press, 1987), 120, 125.

15. Margolis, "Western Coal Mining as a Way of Life," 22.

16. Colorado Fuel and Iron Company, "Statement of Nationality of Employees at Coal Mines and Coke Ovens," August 1, 1918, Company Correspondence, CF&I Archives.

17. Curtis, Fosdick, and Belknap, Report on Industrial Relations in the Colorado Fuel and Iron Company, ca. 1924, n.p., Industrial Relations Materials, CF&I Archives.

18. Colorado Fuel and Iron Company, "Statement of Nationality of Employees."

19. Curtis, Fosdick, and Belknap, Report on Industrial Relations in the Colorado Fuel and Iron Company, n.p.

20. Margolis, "Western Coal Mining as a Way of Life," 44.

21. Paul S. Taylor, *Mexican Labor in the United States*, vol. 1 (Berkeley: University of California Press, 1930), 212.

22. John M. Nieto-Phillips, *The Language of Blood: The Making of Spanish-American Identity in New Mexico, 1880s–1930s* (Albuquerque: University of New Mexico Press, 2004), 4.

23. Paul S. Taylor, *Mexican Labor in the United States Migration: Statistics*, vol. 2 (Berkeley: University of California Press, 1933), 49.

24. Charles Montgomery, *The Spanish Redemption: Heritage, Power, and Loss on New Mexico's Upper Rio Grande* (Berkeley: University of California Press, 2002), 9–10.

25. *The Denver Post*, October 3, 1915.

26. The most distinguishing information on each representative, including his race, comes from the company's personnel card files, which the CF&I Archives has on microfilm. For more on the microfilm collection, see the bibliographic essay at the end of this volume. Not all the information on these cards is discernible. For example, it is sometimes impossible to read the writing on the cards. When there are people with identical names, it is difficult to tell who served as an employee representative as that service is not indicated on his personnel card. Nevertheless, the cards management kept on its workers during this period have a box that describes an employee's race or ethnicity. I have used these company identifications. The managers who filled out the cards had much more information available to come to these judgments. They also had the opportunity to ask the worker to self-identify a characteristic if it was not readily apparent. Therefore, the personnel records are undoubtedly the best source available to identify the race and ethnicity of CF&I's workforce. I am counting "Spanish" and the single use of the classification "mixed" as Mexican American. Other information on race and occupation is from employee representative censuses published in the *Industrial Bulletin* in volume 8 (March 15, 1923, 22–23) and volume 9 (February 15, 1924, 23) and other issues of that publication.

27. Mule drivers, timberers (who did nothing but build the wooden frames that kept the tunnels from collapsing, although some coal diggers had to build their own frames without compensation), and tracklayers who made it possible for the coal carts to travel throughout the mines were shift workers. Fire bosses, who acted as local safety officials, and mine foremen were paid by the day. Ben M. Selekman and Mary Van Kleeck, *Employes' Representation in Coal Mines* (New York: Russell Sage Foundation, 1924), 46.

28. Duane A. Smith, *When Coal Was King: A History of Crested Butte, Colorado, 1880–1952* (Golden: Colorado School of Mines Press, 1984), 79.

29. Ben M. Selekman, "General Statements Concerning Walsen District, Huerfano County," Mary Van Kleeck Papers, Box 47, Folder 6, Sophia Smith Collection, Smith College, Northampton, MA (hereafter Van Kleeck Papers).

30. Jose Villa, Testimony before the Colorado Industrial Commission, "Rest of Walsenburg Covering January 17, 1928," 2091, Industrial Relations Materials, CF&I Archives.

31. Eric Margolis, "Mining Coal in the 1920s: Colorado's Wobbly Strike," in *Slaughter in the Serene: The Columbine Coal Strike Reader*, ed. Lowell May and Richard Myers (Denver: Bread and Roses Cultural Center, 2005), 21.

32. Margolis, "Western Coal Mining as a Way of Life," 45.

33. Colorado Fuel and Iron Company, "Report of the Medical & Sociological Departments of the Colorado Fuel & Iron Company," 1913–1914, 35, 39, Western History Collection, Pueblo Public Library, Pueblo, CO.

34. Thomas G. Andrews, *Killing for Coal: America's Deadliest Labor War* (Cambridge: Harvard University Press, 2008), 175–176.

35. Curtis, Fosdick, and Belknap, Report on Industrial Relations in the Colorado Fuel and Iron Company, 31.

36. Welborn excerpted in George P. West, "Report on the Colorado Strike," in *Massacre at Ludlow: Four Reports*, ed. Leon Stein and Philip Taft (New York: Arno and The New York Times, 1971), 177.

37. Frank L. Palmer, "War in Colorado," *The Nation* 125 (December 7, 1927): 624.

38. Curtis, Fosdick, and Belknap, Report on Industrial Relations in the Colorado Fuel and Iron Company, xi.

39. M. D. Vincent, Testimony before the Colorado Industrial Commission, "Feb. 8–17, 1928," 4044, Industrial Relations Materials, CF&I Archives.

40. Untitled charts, 1928, Box 3b, Folder 19, Mayo Papers.

41. In *Mill & Mine: The CF&I in the Twentieth Century* (Lincoln: University of Nebraska Press, 1992), Lee Scamehorn lists coal mines in operation between 1872 and 1982. It is not comprehensive and has a few errors, but it offers an idea of how fleeting coal mine operations can be. For example, CF&I acquired the Jobal Mine in Huerfano County in 1920. It closed the mine in 1924.

42. Jesse F. Welborn, Testimony before the Colorado Industrial Commission, "Mr. Farrar—Denver, Feb. 18–27, 1928," 3643–44, 3825, Industrial Relations Materials, CF&I Archives. The fact that Welborn could measure the productivity of Mexican American workers meant they were still paid primarily by the number of tons of coal they dug, which therefore means they were still unable to obtain the salaried jobs doing shift work that would have made their employment better paid and more stable.

43. J. F. Welborn to John D. Rockefeller Jr., November 22, 1920, Jesse Floyd Welborn Papers, Box 1, Folder 33, Colorado Historical Society, Denver (hereafter Welborn Papers).

44. Katherine Benton-Cohen, *Borderline Americans: Racial Division and Labor War in the Arizona Borderlands* (Cambridge: Harvard University Press, 2009), 88.

45. *Colorado Fuel and Iron Company Industrial Bulletin* 9 (February 15, 1924): 23.

46. Michael Nuwer, "From Batch to Flow: Production Technology and Work-Force Skills in the Steel Industry, 1880–1920," *Technology and Culture* 29 (October 1988): 837–838, 833.

47. Specific details and more figures relative to this trend throughout the steel industry are found in Michael John Nuwer, "Labor Market Structures in Historical Perspective: A Case Study of Technology and Labor Relations in the United States Iron and Steel Industry," PhD dissertation, University of Utah, 1985, 192–214.

48. James Jones to John D. Rockefeller Jr., April 5, 1935, Rockefeller Family Archives, Record Group III 2 C, Box 13, Folder 109, Rockefeller Archive Center (hereafter RAC), Sleepy Hollow, NY.

49. Mayo to Woods, 32–33, Mayo Papers.

50. Horace B. Drury, "The Three-Shift System in the Steel Industry," *Bulletin of the Taylor Society* 6 (February 1921): 20.

51. *Colorado Fuel and Iron Company Industrial Bulletin* 12 (October 1927): n.p.

52. United States House of Representatives, Subcommittee of the Committee on Mines and Mining, "Conditions in the Coal Mines of Colorado," Part 6 (Washington, DC: Government Printing Office, February 28, March 2, 3, 1914), 1807.

53. Roy E. Dickerson, "SURVEY: Mexican Industrial and Boy Life Pueblo, Colorado," May 13–18, 1921, 11, Box 3b, Folder 18, Mayo Papers.

54. Roy E. Dickerson, "SURVEY: Mexican Industrial Life, C.F.&I. Coal Camps, Colorado," May 19–27, 1921, 7, ibid.

55. E. H. Weitzel to Senator Lawrence C. Phipps, December 17, 1919, Industrial Relations Materials, CF&I Archives.

56. Zaragosa Vargas, *Labor Rights Are Civil Rights: Mexican American Workers in Twentieth Century America* (Princeton: Princeton University Press, 2005), 28–30.

57. Dickerson, "SURVEY: Mexican Industrial Life," 19.

58. Dickerson, "SURVEY: Mexican Industrial and Boy Life," 19.

59. A. T. Jones, "Memo for Mr. Matteson," September 16, 1927, "ERP Reviews," Industrial Relations Materials, CF&I Archives.

60. Curtis, Fosdick, and Belknap, Report on Industrial Relations in the Colorado Fuel and Iron Company, 115.

61. Mayo to Woods, 38, Mayo Papers.

62. Dickerson, "Mexican Industrial and Boy Life," 17.

63. Second Joint Conference—Walsenburg District, May 16, 1919, Box 1, Welborn Papers.

64. B. J. Matteson, "Conferences with Employee Representatives and with the Y.M.C.A. Council and with Negro Employees at Rouse," April 2, 1919, Industrial Relations Materials, CF&I Archives.

65. *Colorado Fuel and Iron Company Industrial Bulletin* 3 (October 31, 1917): 13.

66. Undated Memo, ca. April 1919, "ERP Reviews," CF&I Archives.

67. "William Dow," Mine Personnel Records Reel #1007, CF&I Archives.

68. B. J. Matteson to J. F. Welborn, April 26, 1919, "ERP Reviews," Industrial Relations Materials, CF&I Archives.

69. John D. Rockefeller Jr., "Labor and Capital—Partners," *Atlantic Monthly* (January 1916): 18.

70. Henry A. Atkinson, "The Church and Industrial Warfare," ca. 1914, Rockefeller Family Archives, Record Group III 2 C, Box 20, Folder 186, RAC.

71. Summary of David Griffiths's Reports, June 1916, Industrial Relations Materials, CF&I Archives.

72. Untitled Memo, ca. 1919, Labor Relations Correspondence, ibid.

73. United Mine Workers of America quoted in Paul Nyden, "Miners for Democracy: Struggle in the Coal Fields," PhD dissertation, University of Pittsburgh, 1974, 767.

74. Andrews, *Killing for Coal,* 179.

75. Maier B. Fox, *United We Stand: The United Mine Workers of America, 1890–1990* (Washington, DC: United Mine Workers of America, 1990), 105, 107.

76. John Reed, "The Colorado War," *Metropolitan*, n.d., Special Collections, Tutt Library, Colorado College, Colorado Springs.

77. Edwin Black, *The War against the Weak: Eugenics and America's Campaign to Create a Master Race* (New York: Four Walls Eight Windows, 2003), 93.

78. Quoted in *Colorado Fuel and Iron Company Industrial Bulletin* 1 (January 3, 1916): 3.

79. Fawn-Amber Montoya, "From Mexicans to Citizens: Colorado Fuel and Iron's Representation of Nuevo Mexicans 1901–1919," *Journal of the West* 45 (Fall 2006): 35.

80. See, for example, Daphne Taras, "Voice in the North American Workplace: From Employee Representation to Employee Involvement," in *Industrial Relations to Human Resources and Beyond*, ed. Bruce E. Kaufman, Richard A. Beaumont, and Roy B. Helfgott (Armonk, NY: M. E. Sharpe, 2003), 328, n. 13.

81. Jonathan Rees, *Managing the Mills: Labor Policy in the American Steel Industry during the Nonunion Era* (Lanham, MD: University Press of America, 2003), 190.

82. Selekman and Van Kleeck, *Employes' Representation,* 108.

83. Curtis, Fosdick, and Belknap, Report on Industrial Relations in the Colorado Fuel and Iron Company, 22.

84. "Digest of Main Points of Discussion between Mr. Arthur Young and Dr. E. Mayo," October 15, 1928, 4, Box 3b, Folder 18, Mayo Papers.

85. Hapgood, "Paternalism versus Unionism," 661.

86. Ibid.

87. Ibid.

88. Ibid.

89. Ben M. Selekman, "Report on the Operation of the Agreement between the United Mine Workers of America and the Victor-American Fuel Company," ca. 1919, 41, Box 47, Folder 10, Van Kleeck Papers.

90. Quoted in Raymond B. Fosdick, *John D. Rockefeller Jr.: A Portrait* (New York: Harper & Brothers, 1956), 177–178. That consultant was probably Fosdick himself.

NOTES

CHAPTER 5: THE ROCKEFELLER PLAN IN ACTION: THE MINES

1. Bill Lloyd Interview, May 18, 1978, 10, Eric Margolis Coal Project (Oral History Interviews), Box 10, Archives, University of Colorado at Boulder Libraries, Boulder.

2. John Shaw to E. H. Weitzel, August 13, 1919, Industrial Relations Materials, Colorado Fuel and Iron Archives, Pueblo, CO (hereafter CF&I Archives).

3. E. H. Weitzel to J. F. Welborn, August 15, 1919, ibid.

4. Paul F. Brissenden, "Memorandum," September 12, 1919, ibid.

5. J. F. Welborn to Charles Heydt, October 18, 1919, Rockefeller Family Archives, Record Group III 2 C, Box 15, Folder 127A, Rockefeller Archive Center (hereafter RAC), Sleepy Hollow, NY.

6. W. B. Fitzhugh to C. B. Hudspeth, October 7, 1919, Industrial Relations Materials, CF&I Archives.

7. J. F. Welborn to Martin Radish, September 5, 1919, ibid.

8. John D. Rockefeller Jr. to Mackenzie King, December 30, 1920, Rockefeller Family Archives, Record Group 2, Box 54, Folder 486, RAC. Details of the back and forth between CF&I and the Russell Sage Foundation are contained in subsequent letters from the same file. These studies were not published until January 1925. Van Kleeck revised Selekman's work so substantially that after visiting Colorado herself she became a coauthor of the coal mine study.

9. Edwin Norris to John D. Rockefeller Jr., October 15, 1916, Rockefeller Family Archives, Record Group II 2 C, Box 13, Folder 105, RAC.

10. Thomas G. Andrews, *Killing for Coal: America's Deadliest Labor War* (Cambridge: Harvard University Press, 2008), 168–173.

11. George P. West, "Report on the Colorado Strike," Washington, DC, 1914, 186, in *Massacre at Ludlow: Four Reports*, ed. Leon Stein and Philip Taft (New York: Arno and The New York Times, 1971), 186.

12. Ben M. Selekman and Mary Van Kleeck, *Employes' Representation in Coal Mines* (New York: Russell Sage Foundation, 1924), 150–151.

13. Ibid., 185.

14. David Griffiths to E. S. Cowdrick, July 9, 1918, Industrial Relations Materials, CF&I Archives.

15. C. A. Kaiser to B. J. Matteson, January 19, 1924, ibid.

16. Quoted in Third District Conference, Canon-Western Districts, Rockvale, CO, September 20, 1917, Jesse Floyd Welborn Papers, Box 1, Folder 60, Colorado Historical Society, Denver (hereafter Welborn Papers).

17. Quoted in Third Joint Conference, Trinidad District, September 12, 1922, Box 1, Folder 61, Welborn Papers.

18. B. J. Matteson to A. H. Lichty, June 27, 1922, Industrial Relations Materials, CF&I Archives.

19. John D. Rockefeller Jr., Memorandum Concerning Colorado Trip, 1915, 2, Rockefeller Family Archives, Record Group III 2 C, Box 22, Folder 198, RAC.

NOTES

20. In the 1921 consolidated plan, this body was renamed the Joint Committee on Industrial Cooperation, Conciliation and Wages. The inclusion of the term "wages" in the committee's name in the revised plan reflected the fact that employees' individual wage complaints constituted the committee's primary focus, even though overall wages were supposed to match those at other coal companies back East.

21. Selekman and Van Kleeck, *Employes' Representation*, 429.

22. B. J. Matteson, Testimony before the Colorado Industrial Commission, "Denver Feb. 8–17, 1928," 3879, Industrial Relations Materials, CF&I Archives.

23. H. Lee Scamehorn, *Mill & Mine: The CF&I in the Twentieth Century* (Lincoln: University of Nebraska Press, 1992), 33.

24. "Testimony of E. H. Weitzel Given before the United States Commission on Industrial Relations," Denver, CO, December 8 and 9, 1914, 45, CF&I Archives.

25. James Whiteside, *Regulating Danger: The Struggle for Mine Safety in the Rocky Mountain Coal Industry* (Lincoln: University of Nebraska Press, 1990), xii.

26. Ibid., 125, 94.

27. E. H. Weitzel, December 20, 1915, Colorado Fuel and Iron Collection, Box 12, Colorado Historical Society, Denver.

28. Quoted in "Rest of Walsenburg Covering January 17, 1928," 2196, Industrial Relations Materials, CF&I Archives.

29. "In the Matter of the Death of Cesario Mondragon . . . ," vol. R, Claim #10591, 1921, 86, Commissioners Findings and Awards, Department of Labor and Employment, Colorado Industrial Commission, Colorado State Archives, Denver.

30. Whiteside, *Regulating Danger*, 126.

31. *The Trinidad Evening Picketwire*, March 25, 1922.

32. John Thomas Hogle, "The Rockefeller Plan: Workers, Managers and the Struggle over Unionism in Colorado Fuel and Iron, 1915–1942," PhD dissertation, University of Colorado–Boulder, 1992, 115–116.

33. M. D. Vincent, Testimony before the Colorado Industrial Commission, "Feb. 8–17, 1928," 4051, Industrial Relations Materials, CF&I Archives.

34. Colorado Industrial Commission, "United Mine Workers of America," 1917, 20, Department of Labor and Employment, Colorado Industrial Commission, Box 28617, Colorado State Archives, Denver.

35. Federal Council of Churches, Department of Research and Education, "Industrial Relations in the Coal Industry in Colorado," ca. 1928, 112, Josephine Roche Papers, Box 15, Folder 4, Archives, University of Colorado at Boulder Libraries, Boulder.

36. Hayes quoted in Whiteside, *Regulating Danger*, 121.

37. Scott Martelle, *Blood Passion: The Ludlow Massacre and Class War in the American West* (New Brunswick, NJ: Rutgers University Press, 2007), 19.

38. Selekman and Van Kleeck, *Employes' Representation*, 120.

39. Ben M. Selekman, Interview with George O. Johnson and Harvey Stewart, December 31, 1919, 1–2, Mary Van Kleeck Papers, Box 47, Folder 6, Sophia Smith Collection, Northampton, MA.

40. See, for example, Third Joint Conference, Walsenburg District, Elton Mayo Papers, HBS Archives, Box 3b, Folder 18, Baker Library Historical Collections, Harvard Business School, Boston, MA (hereafter Mayo Papers).

41. Roy E. Dickerson, "SURVEY: Mexican Industrial Life, C.F.&I. Coal Camps, Colorado," May 19–27, 1921, 2, 18, ibid.

42. Colorado Fuel and Iron Company offenses summarized by and quoted in Hogle, "Rockefeller Plan," 123.

43. Report of Adjustments Superintendents with Foremen or Employee Representatives, December 1922, Industrial Relations Materials, CF&I Archives.

44. Report of Adjustments Superintendents with Foremen or Employee Representatives, March 1922, ibid.

45. Summary of David Griffiths's Reports, February 1918, ibid.

46. Digest of Industrial Questions, Box 3b, Folder 19, Mayo Papers.

47. "Rest of Walsenburg," 2242, CF&I Archives.

48. A. Samples to B. J. Matteson, March 1, 1924, Industrial Relations Materials, CF&I Archives.

49. Summary of David Griffiths's Reports, May 1918, ibid.; Selekman and Van Kleeck, *Employes' Representation*, 168–174.

50. Selekman and Van Kleeck, *Employes' Representation*, 175. This case may have been the inspiration for changing the wording in the consolidated plan to make every discharge subject to appeal, as mentioned in Chapter 4.

51. The statute in question is the law that created the Colorado Industrial Commission.

52. United States Senate, Committee on Interstate Commerce, "Increased Price of Coal," Part 3, February 20, 1920, 855.

53. First District Conference, Canon District, January 19, 1917, Box 1, Folder 60, Welborn Papers.

54. First Joint Conference, Canon District, February 24, 1922, Box 1, Folder 61, ibid.

55. *Colorado Fuel and Iron Company Industrial Bulletin* 6 (November 14, 1921): 9.

56. Colorado Industrial Commission, *Fifth Report of the Colorado Industrial Commission December 1, 1920 to December 1, 1921*, 136, Department of Labor and Employment, Colorado Industrial Commission, Box 28617, Colorado State Archives, Denver.

57. Colston E. Warne and Merrill E. Gaddis, "Eleven Years of Compulsory Investigation of Industrial Disputes in Colorado," *Journal of Political Economy* 35 (October 1927): 673.

58. "Wage Readjustments in the Colorado Fuel and Iron Company during 1921," Rockefeller Family Archives, Record Group III 2 C, Box 11, Folder 95, RAC.

NOTES

59. Testimony of Richard Dolan in Official Report of the Proceedings of the National Labor Relations Board, "In the Matter of the Colorado Fuel and Iron Corporation . . . ," vol. 1 (July 5, 1938): 46–51, National Archives II, College Park, MD.

60. Colorado Industrial Commission, *Sixth Report of the Industrial Commission of Colorado December 1, 1921 to December 1, 1922*, 153, Department of Labor and Employment, Colorado Industrial Commission, Box 28617, Colorado State Archives, Denver.

61. *The Pueblo Chieftain*, October 3, 1915.

62. *The Denver Post*, October 3, 1915.

63. *The Pueblo Chieftain*, October 3, 1915.

64. John A. Fitch, "Two Years of the Rockefeller Plan," *The Survey* 39 (October 6, 1917): 14.

65. Ibid., 15.

66. Selekman, Interview with Johnson and Stewart, 3.

67. Colorado Fuel and Iron Company Board of Directors, "Denver, Colorado, February 24, 1916," Minutes of the Meetings of the Board of Directors, vol. B-2, 238, CF&I Archives.

68. [Frank Yaklich Memo], ca. 1925, 34–35, CF&I Archives. [Copy, Unprocessed].

69. Colorado Industrial Commission Statement, 126, Industrial Relations Materials, CF&I Archives.

70. Quoted in First Joint Conference, Trinidad District, January 19, 1922, Box 1, Folder 61, Welborn Papers.

71. Quoted in Third District Conference, Walsenburg District, September 20, 1917, Box 1, Folder 60, Welborn Papers.

72. Starr Murphy to J. F. Welborn, September 18, 1917, Rockefeller Family Archives, Record Group III 2 C, Box 16, Folder 136, RAC.

73. Selekman and Van Kleeck, *Employes' Representation*, 286–288.

74. Ibid., 288.

75. David Brody, *Labor in Crisis: The Steel Strike of 1919* (Urbana: University of Illinois Press, 1987), 129.

76. Selekman and Van Kleeck, *Employes' Representation*, 303–304.

77. J. F. Welborn to All Coal Mine Employees, July 28, 1917, *Colorado Fuel and Iron Company Industrial Bulletin* 2 (July 28, 1917): 13.

78. Donald McClurg, "Labor Organization in the Coal Mines of Colorado, 1878–1930," PhD dissertation, University of California–Berkeley, 1959, 408.

79. Jesse Welborn to J. C. Pacheco, September 8, 1919, Industrial Relations Materials, CF&I Archives.

80. CAK, "Cameron, 8/19/19," ibid.

81. *Colorado Fuel and Iron Company Industrial Bulletin* (January 20, 1920): 4. Selekman and Van Kleeck have pointed out that the request for troops came from only one district, and they even doubt whether it actually originated within the rank and file. See *Employes' Representation*, 315–324.

82. Selekman and Van Kleeck, *Employes' Representation*, 311.
83. Ibid., 340–349.
84. Quoted in United States Senate, Committee on Interstate Commerce, "Increased Price of Coal," 830.
85. Quoted in Selekman and Van Kleeck, *Employes' Representation*, 329.
86. J. F. Welborn to John D. Rockefeller Jr., November 24, 1919, Rockefeller Family Archives, Record Group III 2 C, Box 17, Folder 143, RAC.
87. John D. Rockefeller Jr. to J. F. Welborn, December 2, 1919, ibid.
88. John D. Rockefeller Jr. to A. C. Bedford, November 5, 1919, ibid.

CHAPTER 6: THE ROCKEFELLER PLAN IN ACTION: THE MILL

1. J. F. Chapman to B. J. Matteson, February 14, 1924, Industrial Relations Materials, Colorado Fuel and Iron Archives, Pueblo, CO (hereafter CF&I Archives).
2. Quoted in *The Pueblo Chieftain*, September 26, 1919.
3. John D. Rockefeller Jr., "Representation in Industry," pamphlet, second printing, December 6, 1918, 1.
4. Quoted in *The Pueblo Chieftain*, September 26, 1919.
5. Quoted in *The Denver Post*, November 2, 1919.
6. Quoted in *The Pueblo Chieftain*, September 27, 1919.
7. Ben M. Selekman, *Employes' Representation in Steel Works* (New York: Russell Sage Foundation, 1924), 44, 231.
8. John Thomas Hogle, "The Rockefeller Plan: Workers, Managers and the Struggle over Unionism in Colorado Fuel and Iron, 1915–1942," PhD dissertation, University of Colorado–Boulder, 1992, 257.
9. The definitive book on the fight over the ratification of the Treaty of Versailles and the League of Nations is John Milton Cooper Jr., *Breaking the Heart of the World: Woodrow Wilson and the Fight for the League of Nations* (New York: Cambridge University Press, 2001).
10. Mark M. Jones, Preliminary Report on the Colorado Fuel & Iron Company from the Offices of Curtis, Fosdick, and Belknap, September 1925, n.p., Industrial Relations Materials, CF&I Archives.
11. Ibid.
12. Elton Mayo to Colonel Arthur Woods, November 20, 1928, 26, Elton Mayo Papers, HBS Archives, Box 3b, Folder 18, Baker Library Historical Collections, Harvard Business School, Boston, MA (hereafter Mayo Papers).
13. Report of Adjustments through Superintendents of Foremen with Employee Representatives, July 1924, Industrial Relations Materials, CF&I Archives.
14. Mayo to Woods, 9, Mayo Papers.
15. Official Report of Proceedings of the National Labor Relations Board, "In the Matter of Colorado Fuel and Iron Corporation and Union of Mine, Mill

and Smelter Workers," Cases XXII-R-33 and XX-C-127–134, vol. 6 (July 11, 1938): 646, Administrative Division and Files and Dockets Section, National Labor Relations Board 1935–, Record Group 25, Box 1426, National Archives II, College Park, MD.

16. Selekman, *Employes' Representation*, 228.

17. Minutes of the Employee Representatives Meeting, June 26, 1918, 2–3, Industrial Relations Materials, CF&I Archives.

18. Minutes of the Joint Conference, January 26, 1916, 10, ibid.

19. *Colorado Fuel and Iron Company Industrial Bulletin* 3 (October 31, 1917): 14. The building opened in 1920.

20. Selekman, *Employes' Representation*, 85.

21. Ben M. Selekman and Mary Van Kleeck, *Employes' Representation in Coal Mines* (New York: Russell Sage Foundation, 1924), 241, n. 1. Management granted its employees the eight-hour day shortly before a Colorado law mandating the eight-hour day for coal miners took effect.

22. Horace B. Drury, "The Three-Shift System in the Steel Industry," *Bulletin of the Taylor Society* 6 (February 1921): 19.

23. Ibid.

24. Minutes of the Joint Conference of Employee Representatives for the Minnequa Works, Pueblo, CO, January 19, 1917, Industrial Relations Materials, CF&I Archives.

25. Quoted in Selekman, *Employes' Representation*, 76.

26. Ibid., 76–77.

27. Ibid., 79–81.

28. Ibid., 83.

29. Welborn quoted in "Experience of Colorado Fuel & Iron Co. under 8-Hour Day," *Monthly Labor Review* 17 (August 1923): 405.

30. *Colorado Fuel & Iron Company v. National Labor Relations Board*, 121 F.2d 172 (10th Cir., 1941).

31. Minutes of the Meeting of Employee Representatives, June 29, 1923, *Colorado Fuel and Iron Company Industrial Bulletin* 8 (July 6, 1923): n.p.

32. Minutes of the Joint Conference of Employee Representatives for the Minnequa Works, Pueblo, CO, January 26, 1916, 5–6, Industrial Relations Materials, CF&I Archives.

33. Selekman, *Employes' Representation*, 109, n. 1.

34. Jonathan Rees, *Managing the Mills: Labor Policy in the American Steel Industry during the Nonunion Era* (Lanham, MD: University Press of America, 2004), 88.

35. Selekman, *Employes' Representation*, 98 [extended note 1 from page 97].

36. George W. Zinke, "Causes of Industrial Peace under Collective Bargaining: Minnequa Plant of Colorado Fuel and Iron Corporation and Two Locals of United Steelworkers of America," pamphlet, National Planning Association, Case Study 9, October 1951, 25, CF&I Archives.

37. Selekman, *Employes' Representation*, 109, n. 1.

38. Jones, "Preliminary Report on the Colorado Fuel & Iron Company," n.p.

39. *Colorado Fuel & Iron Company v. National Labor Relations Board*, 121 F.2d 172. Also see "Brief for Respondent, the Colorado Fuel and Iron Corporation," National Labor Relations Board, 1935–, Record Group 25, Administrative Division Files and Dockets Section, Box 1427, National Archives II, College Park, MD.

40. David Brody, *Steelworkers in America: The Nonunion Era* (Cambridge: Harvard University Press, 1960), 268 (reprinted Urbana: University of Illinois Press, 1998).

41. *Chicago Daily Tribune*, September 25, 1915.

42. Curtis, Fosdick, and Belknap, Report on Industrial Relations in the Colorado Fuel and Iron Company, ca. 1924, 53–55, Industrial Relations Materials, CF&I Archives.

43. *Colorado Fuel and Iron Company Industrial Bulletin* 12 (February 1927): 7.

44. B. J. Matteson, "Industrial Relations in the CF&I during 1927," *Colorado Fuel and Iron Company Industrial Bulletin* 13 (January 1928): 12.

45. *Colorado Fuel and Iron Company Industrial Bulletin* 8 (December 15, 1923): 3.

46. Ibid., 5.

47. Ibid., January 15, 1923, 7.

48. Selekman, *Employes' Representation*, 143.

49. Earl Ostrander to Abby Aldrich Rockefeller, March 27, 1929, Rockefeller Family Archives, Business Interests, Box 13, Folder 109, Rockefeller Archive Center (hereafter RAC), Sleepy Hollow, NY.

50. Earl Ostrander (dictated to his mother, Hattie Stricklett) to W. W. Aldrich, April 4, 1934, ibid.

51. No record in the entire Ostrander file at the RAC (Rockefeller Family Archives, Box 13, Folder 109) indicates that management gave Ostrander more money. According to a letter from President Welborn to Rockefeller aide T. M. Debevoise dated April 8, 1929, Ostrander studied law after his accident, was admitted to the bar, and became a county prosecutor.

52. Rees, *Managing the Mills*, 49–50. The classic text on the national steel strike is David Brody, *Labor in Crisis: The Steel Strike of 1919* (Urbana: University of Illinois Press, 1987 [Philadelphia: Lippincott, 1965]).

53. Jesse Welborn to John D. Rockefeller Jr., September 29, 1919, Rockefeller Family Archives, Record Group III 2 C, Box 15, Folder 127A, RAC.

54. Official Report of the Proceedings of the National Labor Relations Board, "In the Matter of Colorado Fuel and Iron Company vs. Union of Mine, Mill and Smelter Workers," Cases XXII-R-33 and XII-C-127–134, vol. 1 (July 5, 1938): 91, Administrative Division and Files and Dockets Section, 1935–, Record Group 25, Box 1426, National Archives II, College Park, MD. Selekman states that 98 percent of the steelworkers voted to join the strike (*Employes' Representation*, 176).

55. J. F. Welborn, "TO OUR EMPLOYEES AND THE PUBLIC," September 19, 1919, Industrial Relations Materials, CF&I Archives.

56. For some unexplained reason, the Colorado Industrial Commission chose not to investigate the 1919 steel strike at CF&I. See Colston E. Warne and Merrill E. Gaddis, "Eleven Years of Compulsory Investigation of Industrial Disputes in Colorado," *Journal of Political Economy* 35 (October 1927): 673.

57. Official Report of Proceedings, vol. 1, 91, National Archives II, College Park, MD.

58. Welborn, "TO OUR EMPLOYEES AND THE PUBLIC," CF&I Archives.

59. Selekman, *Employes' Representation*, 175.

60. Welborn, "TO OUR EMPLOYEES AND THE PUBLIC," CF&I Archives.

61. *The New York Times*, September 22, 1919.

62. Howard M. Gitelman, *Legacy of the Ludlow Massacre: A Chapter in American Industrial Relations* (Philadelphia: University of Pennsylvania Press, 1988), 310.

63. *Pueblo Chieftain*, September 23, 1919.

64. H. Lee Scamehorn, *Mill & Mine: The CF&I in the Twentieth Century* (Lincoln: University of Nebraska Press, 1992), 73.

65. Ibid., 71–72.

66. Quoted in *The Pueblo Chieftain*, October 3, 1915.

67. Gitelman, *Legacy of the Ludlow Massacre*, 310.

68. Quoted in Brody, *Labor in Crisis*, 100.

69. Curtis, Fosdick, and Belknap, in their private study of the plan, freely acknowledged that employee representatives led the back-to-work movement. Management did not admit this to Ben Selekman when he studied the plan. He wrote only that "[m]any intelligent steel workers, who were not union men and not in sympathy with the strike, informed us that in their opinion many steel workers had been stampeded into the Back-to-Work movement, and that they themselves would have nothing to do with it." See Curtis, Fosdick, and Belknap, Report on Industrial Relations at the Colorado Fuel and Iron Company, 19; Selekman, *Employes' Representation*, 187–188.

70. Selekman, *Employes' Representation*, 187–190.

71. William Z. Foster, *The Great Steel Strike and Its Lessons* (New York: B. W. Huebsch, 1920), 186.

72. Colorado Industrial Commission, *Third Report of the Industrial Commission of Colorado December 1, 1918 to December 1, 1919*, 106. Department of Labor and Employment, Colorado Industrial Commission, Box 28617, Colorado State Archives, Denver.

73. Rees, *Managing the Mills*, 190.

74. Jesse F. Welborn, "President Welborn's Report on Strikes in Coal and Steel Operations," *Colorado Fuel and Iron Company Industrial Bulletin* 5 (January 20, 1920): 2.

75. Minutes of the Annual Joint Meeting, Minnequa Works, December 30, 1921, Industrial Relations Materials, CF&I Archives.

76. Second Joint Conference of the Minnequa Works, May 28, 1920, ibid.

77. Minnequa Joint Conference, June 6, 1919, ibid.
78. Ibid.
79. "Joint Committees, December," *Colorado Fuel and Iron Company Industrial Bulletin* 3 (December 17, 1917): n.p.
80. Ben M. Selekman, "Employes' Representation in Steel Works [Draft]," 1923, 121, Rockefeller Family Archives, Record Group III 2 C, Box 16, Folder 135, RAC.
81. Michael John Nuwer, "Labor Market Structures in Historical Perspective: A Case of Technology and Labor Relations in the United States Iron and Steel Industry, 1860–1940," PhD dissertation, University of Utah, 1985, 135.
82. The original version of the steelworks plan is appendix A in Selekman, *Employes' Representation*. For the comparable clause in which this language is absent, see Selekman, 244.
83. Ibid., 127–128.
84. "Questions Presented for Discussion and Adjustment . . . ," March 18, 1921, Industrial Relations Materials, CF&I Archives.
85. Selekman, *Employes' Representation*, 124.
86. Ibid., 211.
87. Ibid., 177.
88. Minutes of the Third Joint Conference of the Minnequa Steel Works, November 29, 1921, Industrial Relations Materials, CF&I Archives.
89. "Joint Conference—Minnequa Works, May 5, 1922," *Colorado Fuel and Iron Company Industrial Bulletin* 7 (May 12, 1922): n.p.
90. Ibid.
91. *Colorado Fuel and Iron Company Industrial Bulletin* 9 (May 15, 1924): n.p.
92. B. J. Matteson to J. F. Welborn, March 29, 1924, Industrial Relations Materials, CF&I Archives.
93. J. F. Welborn, "Seniority of Service," *Colorado Fuel and Iron Company Industrial Bulletin* 12 (April 1927): n.p.

CHAPTER 7: NEW UNION, SAME STRUGGLE

1. Sam Casados to Joe Baros, Box "3–1927," Industrial Relations Materials, Colorado Fuel and Iron Archives, Pueblo, CO (hereafter CF&I Archives). Future researchers should note that during the period of my research for this book, all the materials on the 1927–1928 strike except the minutes of the Colorado Industrial Commission's hearings were boxed together.
2. Richard Myers, "The Columbine Mine Massacre," in *Slaughter in the Serene: The Columbine Coal Strike Reader*, ed. Lowell May and Richard Myers (Denver: Bread and Roses Cultural Center, 2005), 129–131, 138–139.
3. The infamous 1999 Columbine High School massacre in Littleton, Colorado, was, in fact, the second Columbine massacre in Colorado history.
4. *New York Times*, November 24, 1927.

NOTES

5. Myers, "Columbine Mine Massacre," 130, 132.

6. Ronald L. McMahan, "'Rang-u-Tang': The I.W.W. and the 1927 Colorado Coal Strike," in *At the Point of Production: The Local History of the I.W.W.*, ed. Joseph R. Conlin (Westport, CT: Greenwood, 1981), 202.

7. Max Shachtman, "Remember the Ludlow Massacre!" *The Labor Defender* (December 1927): 187.

8. Employee Representatives' Resolution excerpted in "Denver, Feb. 8–17," 3686, Industrial Relations Materials, CF&I Archives.

9. John Thomas Hogle, "The Rockefeller Plan: Workers, Managers and the Struggle over Unionism in Colorado Fuel and Iron, 1915–1942," PhD dissertation, University of Colorado–Boulder, 1992, 178–179.

10. Priscilla Long, *Where the Sun Never Shines: A History of America's Bloody Coal Industry* (New York: Paragon House, 1989), 323.

11. Donald McClurg, "Labor Organization in the Coal Mines of Colorado, 1878–1930," PhD dissertation, University of California–Berkeley, 1959, 442.

12. Committee of Miners to State Industrial Commission, March 29, 1925, Industrial Relations Materials, CF&I Archives.

13. D. A. Stout to Fred Farrar, March 31, 1925, ibid.

14. Hogle, "Rockefeller Plan," 178, 187.

15. Open Letter, September 15, 1927, Industrial Relations Materials, CF&I Archives.

16. E. W. Latchem to Donald McClurg, May 1, 1963, Box 9, Frederick W. Thompson Collection, Walter Reuther Library of Labor and Urban Affairs, Wayne State University, Detroit, MI (hereafter Reuther Library).

17. McMahan, "Rang-u-Tang," 192.

18. E. H. Weitzel to J. F. Welborn, July 19, 1919, Industrial Relations Materials, CF&I Archives.

19. "R" to D. A. Stout, February 8, 1922, ibid.

20. Harry O. Lawson, "The Colorado Coal Strike of 1927–1928," MS thesis, University of Colorado–Boulder, 1950, 52–54.

21. McClurg, "Labor Organization," 450–451.

22. Phil Goodstein, "Colorado's First Columbine Massacre," in *Slaughter in the Serene: The Columbine Coal Strike Reader*, ed. Lowell May and Richard Myers (Denver: Bread and Roses Cultural Center, 2005), 110.

23. Hogle, "Rockefeller Plan," 187.

24. A. S. Embree Statement on untitled IWW flyer, ca. 1927, Josephine Roche Papers, Box 15, Folder 2, Archives, University of Colorado at Boulder Libraries, Boulder (hereafter Roche Papers).

25. McClurg, "Labor Organization," 456–457.

26. Goodstein, "Colorado's First Columbine Massacre," 112.

27. McClurg, "Labor Organization," 461.

28. Hogle, "Rockefeller Plan," 188.

29. Frank L. Palmer, "War in Colorado," *The Nation* 125 (December 7, 1927): 624.

30. Lawson, "Colorado Coal Strike," 203–204.

31. Hogle, "Rockefeller Plan," 189.

32. Vincent quoted in Josephine Roche, "Mines and Men," *The Survey* 61 (December 15, 1928): 342.

33. McClurg, "Labor Organization," 456–457.

34. Kenneth Chorley, "The Labor Situation in Colorado," October 11, 1927, n.p., Rockefeller Family Archives, Record Group III 2 C, Box 14, Folder 121, Rockefeller Archives Center (hereafter RAC), Sleepy Hollow, NY.

35. For more on labor spies at CF&I, see Jonathan Rees, "'XX,' 'XX,' and 'X-3': Spy Reports from the Colorado Fuel and Iron Company Archives," *Colorado Heritage* (Winter 2004): 28–41.

36. Jesse Floyd Welborn, Testimony before the Colorado Industrial Commission, February 28, 1928, 4029, Box 14, Folder 3, Roche Papers.

37. "Joint Meeting of Walsen District Representatives," October 24, 1927, Box "Work Stoppages," CF&I Archives.

38. Welborn, Testimony before the Colorado Industrial Commission.

39. Jesse Floyd Welborn, [Untitled Statement], December 29, 1927, Industrial Relations Materials, CF&I Archives.

40. J. F. Welborn to T. M. Debevoise and Arthur Woods, September 14, 1927, Rockefeller Family Archives, Record Group III 2 C, Box 16, Folder 137, RAC. This sentiment is supported by documentation from late in the strike (January 1928). A list of IWW members that management possessed noted the race, ethnicity, or both of each miner listed "Mexican" as the race for by far the largest number of members. See "List of I.W.W.'s," January 27, 1928, Industrial Relations Materials, CF&I Archives.

41. "Report of XX for Sept. 4, 1927," September 5, 1927, ibid.

42. *Rocky Mountain News,* October 31, 1927.

43. Adams excerpted in McClurg, "Labor Organization," 467.

44. Myers, "Columbine Mine Massacre," 129.

45. Conversation of J. L. McBrayer with Conrad Avillar and John Shepherd, September 8, 1927, Industrial Relations Materials, CF&I Archives.

46. Howard M. Gitelman, *Legacy of the Ludlow Massacre: A Chapter in American Industrial Relations* (Philadelphia: University of Pennsylvania Press, 1988), 338.

47. J. F. Welborn, "President Welborn's Statement of I.W.W. Activities in the Coal Fields," *Colorado Fuel and Iron Company Industrial Bulletin* 12 (December 1927): 3–4.

48. Untitled Meeting Transcript, Ideal Mine, November 8, 1927, 1, Industrial Relations Materials, CF&I Archives.

49. Chorley, "Labor Situation in Colorado," n.p.

50. Lawson, "Colorado Coal Strike," 137.

51. The CF&I Archives has folders of correspondence related to the 1927 strike and many transcripts of these hearings filed separately.

52. Quoted in "Rest of Walsenburg Covering January 17, 1928," Industrial Relations Materials, CF&I Archives.

53. "Testimony of Joe Vosnica, Walsenburg, January 12, 1928," ibid.

54. A. S. Embree, "Bulletin [Transcript]," ca. February 1928, Rockefeller Family Archives, Record Group III 2 C, Box 18, Folder 138, RAC.

55. Industrial Commission of Colorado, "Suggestions and Report of the Industrial Commission of Colorado on the Working Conditions Prevailing in Colorado," March 20, 1927, excerpted in *Colorado Labor Advocate*, March 29, 1928.

56. Industrial Commission of Colorado, "Suggestions and Report of the Industrial Commission of Colorado on the Working Conditions Prevailing in Colorado," March 20, 1927, excerpted in *Colorado Labor Advocate*, April 5, 1928.

57. Ibid.

58. Metal Miners Industrial Union no. 210 of the I.W.W., Bulletin, April 20, 1928, Industrial Workers of the World Papers, Box 51, Reuther Library (hereafter IWW Papers).

59. IWW Educational Bureau, "Coal Mines and Coal Miners: The Story of a Great Industry and the Men Who Work in It," 1923, chapter 18, at Jim Crutchfield's IWW page, http://www.workerseducation.org/crutch/pamphlets/coal/coal.htm, accessed March 11, 2007.

60. UMWA District 15 to Colorado Industrial Commission, 2, Box 73, Folder 73-1, Roche Papers.

61. IWW Educational Bureau, "Coal Mines and Coal Miners," chapter 11.

62. Frank Hefferly to John L. Lewis, December 19, 1924, Box 2, Folder 2-15, Frank and Fred K. Hefferly Papers, Archives, University of Colorado at Boulder Libraries, Boulder.

63. Paul Nyden, "Miners for Democracy: Struggles in the Coal Fields," PhD dissertation, University of Pittsburgh, 1974, 423.

64. McClurg, "Labor Organization," 442–455.

65. McMahan, "Rang-u-Tang," 207.

66. *St. Louis Post Dispatch* quoted in ibid.

67. *Denver Evening News*, October 28, 1927.

68. Roche, "Mines and Men," 343.

69. Mary Van Kleeck, *Miners and Management* (New York: Russell Sage Foundation, 1934), 246.

70. McClurg, "Labor Organization," 541.

71. Coal Miners Industrial Union Bulletin, ca. 1928, Box 15, Folder 15-12, Roche Papers.

72. "Agreement by and between the Rocky Mountain Fuel Company and the United Mine Workers of America," District no. 15, September 1, 1928, 3, ibid.

73. Rees, "'X,' 'XX,' and 'X-3,'" 41.

74. Metal Miners Industrial Union no. 210 of the I.W.W., Bulletin, April 20, 1928, Box 51, IWW Papers.

75. Edward Costigan to Josephine Roche, "Spring 1929," 8, Box 73, Folder 73-1, Roche Papers.

NOTES

76. *New York Times*, September 23, 1928.

77. Barron B. Beshoar, *Out of the Depths: The Story of John R. Lawson, a Labor Leader* (Denver: Golden Bell, 1942), 365.

78. Rocky Mountain Fuel Company, "A Challenge to Organized Labor!" Undated Flier, Box 17, Folder 17-5, Roche Papers.

79. Goodstein, "Colorado's First Columbine Massacre," 123.

80. *The New York Times*, August 27, 1931.

81. Untitled Notice, January 10, 1928, Elton Mayo Papers, HBS Archives, Box 3b, Folder 18, Baker Library Historical Collections, Harvard Business School, Boston, MA.

82. H. Lee Scamehorn, *Mill & Mine: The CF&I in the Twentieth Century* (Lincoln: University of Nebraska Press, 1992), 133.

83. J. F. Welborn to Clarence Hicks, May 28, 1932, Rockefeller Family Archives, Record Group III 2 C, Box 14, Folder 121, RAC.

84. G. H. Rupp to Arthur Roeder and W. A. Maxwell, July 11, 1933, Industrial Relations Materials, CF&I Archives.

85. Ibid., 2.

86. *CF&I Blast*, December 16, 1932.

87. Zaragosa Vargas, *Labor Rights Are Civil Rights: Mexican American Workers in Twentieth Century America* (Princeton: Princeton University Press, 2005), 51.

88. Arthur Roeder to Colorado State Industrial Commission, June 1, 1931, Box 1, Folder 1-16, Colorado State Federation of Labor Papers, Archives, University of Colorado at Boulder Libraries, Boulder (hereafter Colorado State Federation of Labor Papers).

89. Van Kleeck, *Miners and Management*, 120–129.

90. Tony Fatur et al. to Colorado State Industrial Commission, June 13, 1931, Box 1, Folder 1-14, Colorado State Federation of Labor Papers.

91. Colorado Industrial Commission quoted in Van Kleeck, *Miners and Management*, 125.

92. Josephine Roche to John D. Rockefeller Jr., August 1, 1931, Box 17, Folder 17-9, Roche Papers.

93. Quoted in *The New York Times*, August 3, 1931.

94. Van Kleeck, *Miners and Management*, 134.

95. The change in policy might also be attributed in part to the company's slippage into receivership, which meant its creditors had a say over any policy that might affect its earnings. See Hogle, "Rockefeller Plan," 215–216.

96. Ibid., 215.

97. McClurg, "Labor Organization," 575–576.

98. Hogle, "Rockefeller Plan," 214–216.

99. Arthur Roeder excerpted in Van Kleeck, *Miners and Management*, 324–325.

100. John D. Rockefeller Jr. to Mackenzie King, August 31, 1933, Rockefeller Family Archives, Record Group 2, Box 9, Folder 86, RAC.

101. Scamehorn, *Mill & Mine*, 147.

NOTES

CHAPTER 8: DEPRESSION, FRUSTRATION, AND REAL COMPETITION

1. Andrew Diamond to John D. Rockefeller Jr., August 23, 1933, Rockefeller Family Archives, Record Group III 2 C, Box 14, Folder 14, Rockefeller Archive Center (hereafter RAC), Sleepy Hollow, NY.

2. *Pueblo Chieftain*, June 8, 1926.

3. Quoted in "New Business," *Colorado Fuel and Iron Company Industrial Bulletin* (July 28, 1926): n.p.

4. David Rockefeller, *Memoirs* (New York: Random House, 2002), 42.

5. John D. Rockefeller Jr. to Mackenzie King, September 28, 1928, in JDR Jr. Personal, Record Group III 2 Z, Box 54, Folder 486, RAC.

6. John D. Rockefeller Jr. to John D. Rockefeller III, April 15, 1929, quoted in John Ensor Harr and Peter J. Johnson, *The Rockefeller Century* (New York: Charles Scribner's Sons, 1988), 272–273.

7. John D. Rockefeller III, "Industrial Relations Plans: A Study," senior thesis, Princeton University, 1929, 18–19, Seeley G. Mudd Library, Princeton, NJ.

8. *Pueblo Chieftain*, July 31, 1933.

9. H. Lee Scamehorn, *Mill & Mine: The CF&I in the Twentieth Century* (Lincoln: University of Nebraska Press, 1992), 134; Jonathan Rees, *Managing the Mills: Labor Policy in the American Steel Industry during the Nonunion Era* (Lanham, MD: University Press of America, 2004), 167.

10. Elton Mayo to Colonel Arthur Woods, November 20, 1928, 9, Elton Mayo Papers, HBS Archives, Box 3b, Folder 19, Baker Library Historical Collections, Harvard Business School, Boston, MA.

11. Third Joint Conference, Minnequa Steel Works, September 28, 1928, 4, ibid.

12. John Thomas Hogle, "The Rockefeller Plan: Workers, Managers and the Struggle over Unionism in Colorado Fuel and Iron, 1915–1942," PhD dissertation, University of Colorado–Boulder, 284.

13. Annual Joint Meeting, Minnequa Steel Works, December 10, 1931, 9, Industrial Relations Materials, Colorado Fuel and Iron Archives, Pueblo, CO (hereafter CF&I Archives).

14. Scamehorn, *Mill & Mine*, 138.

15. Annual Joint Meeting, Minnequa Steel Works, December 10, 1931, 7, Industrial Relations Materials, CF&I Archives.

16. Scamehorn, *Mill & Mine*, 145–146, 138–139.

17. *Pueblo Chieftain*, August 11, 1933.

18. Annual Joint Meeting, Minnequa Steel Works, December 12, 1935, Industrial Relations Materials, CF&I Archives.

19. "Restlessness in Steel," *Fortune* 8 (September 1933): 122.

20. *New York Times*, August 2, 1933.

21. Hogle, "Rockefeller Plan," 286. On the remaining mines, see Scamehorn, *Mill & Mine,* 201–203. The only mines he lists that the company owned when the

Rockefeller Plan was implemented in 1915 and that had not closed prior to 1936 were Frederick, Pictou, Cameron, Kebler No. 2, and Rockvale No. 5.

22. Curtis, Fosdick, and Belknap, Report on Industrial Relations in the Colorado Fuel and Iron Company, 1925, 21–22, Industrial Relations Materials, CF&I Archives.

23. Mark M. Jones, "Preliminary Report on the Colorado Fuel & Iron Company from the Offices of Curtis, Fosdick and Belknap," September 1925, n.p., ibid.

24. Quoted in *The Denver Post*, July 19, 1933.

25. *CF&I Blast*, October 31, 1930.

26. Quoted in *Colorado Fuel and Iron Company Industrial Bulletin* 15 (November–December 1930): 5.

27. In 1928 the budget deficit at the YMCA was $58,500. Even when the building was mostly boarded up in 1935, the deficit was still $15,000. See Arthur Roeder to Arthur Packard, November 30, 1936, Rockefeller Family Archives, Record Group III 2 C, Box 18, Folder 164, RAC.

28. *Pueblo Chieftain*, August 11, 1933.

29. Nelson Rockefeller to Arthur Roeder, August 30, 1933, Rockefeller Family Archives, Record Group III 2 C, Box 14, Folder 114, RAC.

30. Quoted in *CF&I Blast*, March 24, 1933.

31. Quoted in ibid., March 31, 1933.

32. Ibid., April 14, 1933.

33. Arthur Roeder to Thomas M. Debevoise, October 20, 1936, Rockefeller Family Archives, Record Group III 2 C, Box 18, Folder 164, RAC.

34. Diamond to Rockefeller, August 23, 1933, RAC.

35. In 1937 CF&I bore most of the expense to reopen the "Y," with much fanfare, as part of its campaign against the newly formed Steel Workers Organizing Committee. It closed again sometime in the 1950s.

36. Arthur Roeder to Arthur Woods, August 21, 1933, Rockefeller Family Archives, Record Group III 2 C, Box 14, Folder 114, RAC.

37. *Pueblo Chieftain*, August 11, 1933.

38. Robert T. Gumbel to Andrew J. Diamond, ca. August 1933, Rockefeller Family Archives, Business Interests, Box 114, Folder 14, RAC.

39. Lawson quoted in F. Darrell Munsell, *From Redstone to Ludlow: John Cleveland Osgood's Struggle against the United Mine Workers of America* (Boulder: University Press of Colorado, 2009), 353. This phenomenon is typical of many ERPs. As Greg Patmore suggested after reviewing employee representation plans in three countries during this era, "[T]he presence of these plans can be a question of management style as much as a reflection of the economic climate." See Patmore, "Employee Representation Plans in the United States, Canada and Australia: An Employer Response to Workplace Democracy," *Labor: Studies in Working Class History of the Americas* 3 (Summer 2006): 59.

40. Lizabeth Cohen, *Making a New Deal: Industrial Workers in Chicago, 1919–1939* (New York: Cambridge University Press, 1990), 246.

NOTES

41. Quoted in Official Report of the Proceedings before the National Labor Relations Board, "In the Matter of the Colorado Fuel and Iron Corporation and International Union of Mine, Mill and Smelter Workers," 633, Records of the United States Court of Appeals, Tenth Circuit, Box 202, Record Group 276, National Archives and Records Administration, Rocky Mountain Region, Lakewood, CO (hereafter Records of the United States Court of Appeals, Tenth Circuit).

42. "Anderson, James," CF&I Company Personnel Records, Reel 1023, Aaby–Julio Carillo, CF&I Archives.

43. Quoted in *Colorado Fuel and Iron Company Industrial Bulletin* 29 (August–December 1934): 8.

44. Ibid., 20 (February–May 1935): 12.

45. National Labor Relations Board, Official Report of Proceedings, National Labor Relations Board 1935–, Administrative Division, Box 1427, Record Group 25, National Archives II, College Park, MD (hereafter National Archives II).

46. *Colorado Fuel and Iron Company Industrial Bulletin* 20 (May–June 1935): 9–19; Hogle, "Rockefeller Plan," 291–292.

47. Quoted in *Colorado Fuel and Iron Company Industrial Bulletin* 20 (December 1934–February 1935): 4, 13.

48. Official Report of the Proceedings before the National Labor Relations Board, "In the Matter of the Colorado Fuel and Iron Corporation and International Union of Mine, Mill and Smelter Workers," 44, Records of the United States Court of Appeals, Tenth Circuit. Even though the company offered Anderson another job, he did not accept it. See CF&I Personnel Records, Reel 1023, CF&I Archives, Pueblo.

49. Quoted in *Colorado Fuel and Iron Company Industrial Bulletin* 20 (February–May 1935): 22.

50. Ibid., 29 (March–May 1934): 26, 30.

51. "In the Matter of the Colorado Fuel and Iron Corporation and Steel Workers Organizing Committee," Brief for the Respondent, the Colorado Fuel and Iron Corporation, n.d., 21, National Labor Relations Board 1935–, Administrative Division Files, Box 1427, National Archives II.

52. *CF&I Blast*, June 23, 1933.

53. Rees, *Managing the Mills*, 191.

54. "Characteristics of Company Unions 1935," United States Department of Labor, Bureau of Labor Statistics, Bulletin 634, June 1937, 204.

55. David J. Saposs, "Organizational and Procedural Changes in Employee Representation Plans," *Journal of Political Economy* 44 (December 1936): 807–808.

56. John Pencavel, "Company Unions, Wages and Work Hours," *Advances in Industrial and Labor Relations* 12 (Amsterdam: JAI, 2003): 23.

57. National Labor Relations Board, "National Labor Relations Act," http://www.nlrb.gov/about_us/overview/national_labor_relations_act.aspx, accessed April 20, 2007.

NOTES

58. James D. Rose, *Duquesne and the Rise of Steel Unionism* (Urbana: University of Illinois Press, 2001), 138.

59. Rees, *Managing the Mills*, 220–221.

60. For example, see "INGRESE A LA-UNION," Undated Flier, Box 2, Folder 2-31, Frank and Fred K. Hefferly Papers, Archives, University of Colorado at Boulder Libraries, Boulder (hereafter Hefferly Papers).

61. *Colorado Fuel and Iron Company Industrial Bulletin* 21 (January–April 1936): 25. These changes came about as a result of the passage of the NLRA and the desire for employee representatives to mandate arbitration following the Anderson case. Because of restrictions imposed by the NLRA, the committee making the revisions consisted only of employee representatives. The most important changes in the new version of the plan included having all employee representative elections conducted by a special committee of five employee representatives rather than by management, a provision for monthly (rather than quarterly) meetings of employee representatives, the possibility of outside arbitration by a board of three that would include one outside party mutually agreed upon by both sides, and new provisions opening the possibility of amending or terminating the plan. The revisions also cut management's financial support for the plan, although it did not eliminate it. For details see 22 N.L.R.B., no. 14, 1940, 200–205.

62. Management had every reason to hide its involvement in this set of revisions. While I have found no documentation to prove that it initiated these changes, if CF&I resembled similar companies at the time, management probably did so. *NLRB v. Jones and Laughlin Steel Corp.*, 301 U.S. 1 (1937).

63. This version of the Rockefeller Plan further cut financial support by management for the operation of the ERP, forcing company union members to pay dues. It also deleted all language referring to the joint nature of the plan and included language in the Memorandum of Agreement that implied that management had a role in all these policies.

64. 22 N.L.R.B., no. 14, 1940, 205–208, 17.

65. Hogle, "Rockefeller Plan," 299.

66. Scamehorn, *Mill & Mine*, 149–150.

67. "In the Matter of the Colorado Fuel and Iron Corporation and Steel Workers Organizing Committee," Case no. R-2190, 1940, 3, n. 3, Box 203, Records of the United States Court of Appeals, Tenth Circuit.

68. Joint Report of the Executive Officers and Executive Board of the Colorado State Industrial Union Council CIO, Pueblo, CO, September 20, 1940, 6, Box 2, Folder 2-31, Hefferly Papers.

69. National Labor Relations Board, Official Report of Proceedings, "In the Matter of Colorado Fuel and Iron Corporation and Union of Mine, Mill and Smelter Workers," vol. 1 (July 5, 1938): 28–29, National Labor Relations Board, Administrative Division, 1935–, Box 1426, National Archives II.

70. National Labor Relations Board, Official Report of Proceedings, "In the Matter of the Colorado Fuel and Iron Corporation and International Union of

Mine, Mill and Smelter Workers," 128–129, Records of the United States Court of Appeals, Tenth Circuit.

71. Steel Workers Organizing Committee, "Take Off Your Mask, Mr. Diamond," Undated Flier, Box 4, Folder 4-15, Hefferly Papers. Diamond's middle initial was J, probably for Jack or a name shortened to Jack.

72. Steel Workers Organizing Committee, "Why Support a Company Union?" Undated Flier, ibid.

73. Steel Workers Organizing Committee, "The Big Sideshow," Undated Flier, ibid.

74. Hogle, "Rockefeller Plan," 299.

75. National Labor Relations Board, Official Report of Proceedings, August 31, 1939, National Labor Relations Board, Administrative Division, 1935–, Box 1427, National Archives II.

76. Nobody discussed whether management paid for Diamond and Irwin's travel expenses, but since management regularly paid for employee representatives to survey wages in eastern mills, this is a safe assumption.

77. 22 N.L.R.B., no. 14, 1940, 218–219, 212.

78. Despite objections from CF&I, the two cases were merged shortly after the two unions filed their complaints in March 1938. The Sunrise, Wyoming, case became the lead dispute, even though it affected far fewer people than the one in Pueblo.

79. *The Pueblo Chieftain*, March 2, 1941.

80. "In the Matter of the Colorado Fuel & Iron Corporation and Steel Workers Organizing Committee," Case no. R-2190, 1940, Supplemental Decision and Orders, *Colorado Fuel and Iron Company vs. National Labor Relations Board*, Box 201, Records of the United States Court of Appeals, Tenth Circuit.

81. Hogle, "Rockefeller Plan," 304.

82. *Pueblo Chieftain*, March 2, 1941.

83. Frank Bonacci, "Report of Frank Bonacci," March 22, 1941, Box 2, Folder 2-31, Hefferly Papers. SWOC's complacent attitude toward the election can be seen in a radio address Frank Hefferly gave before the vote. "All indications point to a sweeping victory by SWOC," he declared, "notwithstanding loud noise to the contrary made by leaders of the ill-fated Employee Representative Plan and other enemies." Frank Hefferly, "Steel Workers of Pueblo," ca. March 1941, Box 2, Folder 2-31, Hefferly Papers.

84. Undated, unlabeled newspaper clipping, Box 7, Folder 7-15, Hefferly Papers.

85. *Colorado Fuel & Iron Corporation v. National Labor Relations Board*, 121 F.2d 165 (10th Cir., 1941), 174.

86. "Meeting Held at the Colorado Fuel and Iron Corporation's Y.M.C.A. Building . . . ," July 21, 1941, Box 201, Records of the United States Court of Appeals, Tenth Circuit.

87. Hogle, "Rockefeller Plan," 306–307.

88. "Agreement between the Colorado Fuel and Iron Corporation and Employee Representatives Organization, a Corporation," December 1, 1941, Industrial Relations Materials, CF&I Archives.

89. Hogle, "Rockefeller Plan," 308–309.

90. Quoted in "In the Matter of the Colorado Fuel & Iron Corporation and Steel Workers Organizing Committee," 38, Records of the United States Court of Appeals, Tenth Circuit.

91. Rees, *Managing the Mills*, 265.

92. "In the Matter of the Colorado Fuel & Iron Corporation and United Steelworkers of America," N.L.R.B., no. 43 (1940), 300.

93. Hogle, "Rockefeller Plan," 312.

94. Many thanks to Joe Koncilja for giving me one of the knives.

95. *Rocky Mountain News*, May 15, 2004.

96. Rockefeller quoted in Daniel Okrent, *Great Fortune: The Epic of Rockefeller Center* (New York: Penguin, 2003), 390.

97. Ibid., 391.

98. Quoted in *The Pueblo Chieftain*, December 24, 1944.

CONCLUSION

1. Robert Wagner, "Company Unions: A Vast Industrial Issue," *The New York Times*, March 11, 1934.

2. Howard Briggs to John D. Rockefeller Jr., December 23, 1949, Rockefeller Family Archives, Record Group III 2 C, Box 19, Folder 166, Rockefeller Archive Center, Sleepy Hollow, NY.

3. Daniel Okrent, *Great Fortune: The Epic of Rockefeller Center* (New York: Penguin, 2003), 132.

4. Gary Dean Best, "President Wilson's Second Industrial Conference, 1919–20," *Labor History* 16 (Fall 1975): 520.

5. David Brody, "Section 8(a)(2) and the Origins of the Wagner Act," in *Labor Embattled: History, Power, Rights* (Urbana: University of Illinois Press, 2005), 51.

6. I make this argument because I heard historian Pauline Maier make the same argument with respect to the term "Anti-Federalists" at a speech preceding the publication of her forthcoming study of the ratification of the Constitution.

7. John Pencavel, "Company Unions, Wages, and Work Hours," *Advances in Industrial and Labor Relations*, vol. 12, ed. David Lewin and Bruce E. Kaufman (Amsterdam: JAI, 2003): 31.

8. I am borrowing an argument from Bruce E. Kaufman and Daphne Gottlieb Taras, "Conclusion," in *Nonunion Employee Representation: History, Contemporary Policy, Practice*, ed. Bruce E. Kaufman and Daphne Gottlieb Taras (Armonk, NY: M. E. Sharpe, 2000), 547.

NOTES

9. Brody, "Section 8(a)(2)," 47.

10. *Texas and New Orleans Railroad Company v. Brotherhood of Railway and Steamship Clerks*, 281 U.S. 548 (1930).

11. National Industrial Recovery Act, "Our Documents," National Archives and Records Administration, http://www.ourdocuments.gov/doc.php?flash=true &doc=66&page=transcript, accessed December 29, 2007.

12. Since the National Industrial Recovery Act had been completely invalidated by the U.S. Supreme Court in early 1935, Section 7(a) of the NIRA had to be replaced.

13. For example, Section 9(a) reads, in part, "Representatives designated or selected for the purposes of collective bargaining by the majority of the employees in a unit appropriate for such purposes, shall be the exclusive representatives of all the employees in such unit for the purposes of collective bargaining in respect to rates of pay, wages, hours of employment, or other conditions of employment." See National Labor Relations Act, "Our Documents," National Archives and Records Administration, http://www.ourdocuments.gov/doc.php?doc=67&page =transcript, accessed February 22, 2008.

14. Raymond L. Hogler, "Worker Participation, Employer Anti-Unionism, and Labor Law: The Case of the Steel Industry, 1918–1937," *Hofstra Labor Law Journal* 7 (Fall 1989): 39–41. Hogler quoted the comment about an open field from a compendium of the legislative history on the NLRA.

15. Wagner quoted in Raymond L. Hogler, "Exclusive Representation and the Wagner Act: The Structure of Federal Collective Bargaining Law," *Labor Law Journal* 58 (Fall 2007): 162.

16. *National Labor Relations Board v. Newport News Shipbuilding & Dry Dock Co.*, 308 U.S. (1939), 241.

17. Sanford M. Jacoby, "A Road Not Taken: Independent Local Unions in the United States since 1935," in *Nonunion Employee Representation: History, Contemporary Policy, Practice*, ed. Bruce E. Kaufman and Daphne Gottlieb Taras (Armonk, NY: M. E. Sharpe, 2000), 78.

18. Charles J. Morris, *The Blue Eagle at Work: Reclaiming Democratic Rights in the American Workplace* (Ithaca, NY: ILR Press, 2005), 6.

19. David Brody, "Labor Elections: Good for Workers?" *Dissent* 44 (Summer 1997): 71–77.

20. Thomas Geoghegan, "Taking It to the Blue States," *The Nation*, November 29, 2004, http://www.thenation.com/docprem.mhtml?i=20041129&s=geog hegan, accessed November 30, 2005.

21. *Electromation, Inc. and International Brotherhood of Teamsters, Local Union No. 149 . . .* , 309 N.L.R.B. 990 (1992), 1003.

22. Ibid.

23. *E. I. du Pont de Nemours & Company and Chemical Workers Association, Inc. . . .* , 311 N.L.R.B. 893 (1993).

24. Had these decisions been enforced since the early days of the NLRA, no "company unions" could have survived the onslaught. However, as Sanford

Jacoby pointed out in "Current Prospects for Employee Representation in the U.S.: Old Wine in New Bottles?" (*Journal of Labor Research* 16 [Summer 1995]: 387–398), many did survive by both accident and design.

25. "The Teamwork for Employees and Managers (TEAM) Act," excerpted in John W. Budd, *Labor Relations: Striking a Balance* (Boston: McGraw-Hill, 2005), 387.

26. Jonathan P. Hiatt and Laurence E. Gold, "Employer-Employee Committees: A Union Perspective," in *Nonunion Employee Representation: History, Contemporary Policy, Practice*, ed. Bruce E. Kaufman and Daphne Gottlieb Taras (Armonk, NY: M. E. Sharpe, 2000), 508.

27. U.S. Commission on the Future of Worker-Management Relations, "Final Report," 1994, http://digitalcommons.ilr.cornell.edu/key_workplace/2/, accessed May 14, 2007.

28. Nelson Lichtenstein, *State of the Union* (Princeton, NJ: Princeton University Press, 2002), 245.

29. Kaufman quoted in Pencavel, "Company Unions," 30.

30. Bruce E. Kaufman, "Does the NLRA Constrain Employee Involvement and Participation Programs in Nonunion Companies? A Reassessment," *Yale Law and Policy Review* 17, 2 (1999): 729–811.

31. Raymond L. Hogler and Guillermo J. Grenier, *Employee Participation and Labor Law in the American Workplace* (New York: Quorum Books, 1992), 146.

32. John Logan, "The Debate over Employee Representation Plans and Works Councils during the Clinton Administration," 2–3, Symposium on Non-Union Forms of Employee Representation in the Asia Pacific Rim, University of Sydney, Sydney, Australia, December 7, 2007.

33. Jacoby, "Current Prospects," 391–392.

34. Quoted in *The New York Times*, October 1, 1915.

35. See, for example, Thomas Frank, *One Market under God: Extreme Capitalism, Market Populism and the End of Economic Democracy* (New York: Random House, 2000), 179.

36. Lichtenstein, *State of the Union*, 148–149.

37. Ian Sakinofsky, "Consultation in the Workplace (between Superior and Subordinate) as a Non-Equivalent Substitute for True Worker Voice, and as a Means of Securing Worker Compliance with Unpopular Decisions," 4, Symposium on Non-Union Forms of Employee Representation in the Asia Pacific Rim.

38. Hogler and Grenier, *Employee Participation and Labor Law*, 119.

39. Franklin D. Roosevelt, "Excerpts from the Press Conference," June 15, 1934, The American Presidency Project, University of California at Santa Barbara, http://www.presidency.ucsb.edu/ws/print.php?pid=14694, accessed December 20, 2007.

40. Reg Basken, "My Experience with Unionization of Nonunion Employee Representation Plans in Canada," in *Nonunion Employee Representation: History, Contemporary Policy, Practice*, ed. Bruce E. Kaufman and Daphne Gottlieb Taras (Armonk, NY: M. E. Sharpe, 2000), 487.

41. Rae Cooper and Chris Briggs, "'Trojan Horse' or 'Vehicle for Organising'? Nonunion Collective Agreement Making and Trade Unions in Australia," 21, Symposium on Non-Union Forms of Employee Representation in the Asia Pacific Rim.

42. Wolfgang Streeck, "Co-determination: After Four Decades," in *Social Institutions and Economic Performance*, ed. Wolfgang Streeck (London: Sage, 1992): 153. *Works Councils: Consultation, Representation, and Cooperation in Industrial Relations*, ed. Joel Rogers and Wolfgang Streeck (Chicago: University of Chicago Press, 1995), contains many other examples from countries across Europe as further proof that non-union employee representation can be advantageous to independent unions.

BIBLIOGRAPHIC ESSAY

The Colorado Fuel and Iron Company (CF&I) Archives is undoubtedly one of the largest corporate archives available to anyone interested in studying the history of a U.S. business. As I write these words, it is also in disarray. A team of archivists was working on putting this huge collection in order during the three-year period in which I researched and wrote this book. After a long period with no archivists, the Bessemer Historical Society recently received another grant from the National Endowment for the Humanities to fund salaries for a new team. It is hoped that these positions will prove permanent as there is much work left to do.

The vast majority of the material used in this study is part of a record group now called the Industrial Relations Record Group. A description of the record group (and the rest of the CF&I Archives) is available online at the Rocky Mountain Online Archive, http://rmoa.unm.edu/. Click on the tab for the Bessemer Historical Society. Current box and folder descriptions should be available there. However, since this collection was unprocessed when I used it, in my notes I used the name the record group had when I did most of my research: Industrial Relations Materials. Since box numbers and folder names have changed several times since I started using the archives and may change again as more material is discovered, I refer to each record from this group only by its name and this record group. It is hoped that further processing will make

BIBLIOGRAPHIC ESSAY

these records better organized and more accessible in the years following the publication of this book.

The Rockefeller Plan meeting minutes are currently intermixed in these records with various other reports and correspondence related to the employee representation plan (ERP). The minutes I have seen include meetings from every period in the plan's history from the 1910s to the early 1940s. The quality of the meeting minutes varies greatly. Some are exact transcripts of what everyone present (both labor and management) said during the meeting. Others are detailed summaries written from the fly-on-the-wall perspective of an unnamed management stenographer.[1] The most voluminous minutes are the shorter summaries management assembled for publication in the *Colorado Fuel and Iron Company Industrial Bulletin*.

The CF&I Archives has two complete sets of the published *Industrial Bulletin*, and it is also available at many other libraries across the United States. The only set of supplemental minutes containing the transcripts of meetings from the 1930s I have found is at the Denver branch of the National Archives, in Lakewood, Colorado, housed with the records of the 1940 court case that challenged the establishment of the plan by the National Labor Relations Board. The *Industrial Bulletin* is also a very important source because of its regular reporting, not just on welfare capitalism but also on important events in the company's history, such as strikes.

Retired steelworkers told Bessemer Historical Society archivists that the original personnel files were destroyed when they were first microfilmed during the 1940s. BHS has made usable copies of those microfilms, but they are far from complete and are completely disorganized. Most notably, they are not all organized alphabetically and are not numbered by frame. Therefore, it is likely that my research assistants and I may have missed a card somewhere in that sea of microfilm.

In addition to the material in Pueblo, I used a considerable amount of documentation from the Rockefeller Archive Center in Sleepy Hollow, New York, relating to both the Rockefeller Plan and John D. Rockefeller

---

1. The names given to each summary are inconsistent. The titles of minutes for different levels of meetings range from two or three words to long descriptions. In each instance where I quote such a document, I simply title it using whatever is at the top of the first page.

BIBLIOGRAPHIC ESSAY

Jr.[2] During my limited research time there, I concentrated on documents directly related to Colorado Fuel and Iron. For broader Rockefeller-related archival material, I relied primarily on documents cited in the work of other authors. The papers of CF&I president Jesse Floyd Welborn at the Colorado Historical Society in Denver include many earlier Rockefeller Plan meeting minutes from coal mines across the state that have not been found at the CF&I Archives. The Elton Mayo Papers at the Harvard Business School's Baker Library include extensive documentation of Mayo's trip to Colorado in 1928 to see the plan in operation. Another contemporary source of information on the Rockefeller Plan that remains extremely useful is Ben M. Selekman and Mary Van Kleeck's landmark 1924 study for the Russell Sage Foundation, *Employes' Representation in Coal Mines,* and its companion volume, *Employes' Representation in Steel Works,* by Selekman alone. Both works were based on extensive fieldwork in Colorado during 1919 and 1921.

Historians have not treated the Rockefeller Plan kindly, and there are good reasons for this attitude. While my position on the plan is very different, Howard Gitelman's *Legacy of the Ludlow Massacre* is the best previously published work that considers the plan at any length. John Thomas Hogle's unpublished 1992 University of Colorado dissertation, "The Rockefeller Plan: Workers, Managers and the Struggle over Unionism in Colorado Fuel and Iron, 1915–1942," is also a fine work. However, the analysis is handicapped by the fact that the vast Colorado Fuel and Iron Archives, including the minutes of the ERP meetings, was not open when he wrote it.

The best books I have found on the Rockefeller family are Ron Chernow's *Titan,* Peter Collier and David Horowitz's *The Rockefellers,* and John Ensor Harr and Peter J. Johnson's *The Rockefeller Century.* Although it contains little about Colorado, Daniel Okrent's *Great Fortune: The Epic of Rockefeller Center* sheds significant light on John D. Rockefeller Jr.'s character. As I note in Chapter 2, Raymond Fosdick's *John D. Rockefeller, Jr.: A Portrait* is too hagiographic to be particularly useful, but no other biography of the younger Rockefeller exists.

---

2. This includes both sides of his voluminous correspondence with Mackenzie King. The Rockefeller Archive Center has copies of Rockefeller's letters to King, which saved me a trip to Ottawa to see the originals.

INDEX

Page numbers in italics indicate illustrations.

Accidents, 97, 118; Minnequa Works, 145–46, 147, *148*
Adams, William H., 167, 168
Advisory Board, under Rockefeller Plan, 233–34
AFL. *See* American Federation of Labor
African Americans, 89, 101–2
Aguilar, strike at, 163, 164–65
Aldrich, Nelson, 147
Allen, Charles, Jr., 205
Amalgamated Association of Iron, Steel and Tin Plate Workers, 85, 107–8, 141; 1919 strike, 149–52
American Federation of Labor (AFL), 7, 29, 63, 68, 174, 175, 193
Anderson, James, dismissal of, 190–91, 199–200
Anti-unionism, 2, 6, 8, 29–30, 54
Atkinson, Henry A., 103
Australia, 219
Austrians, 90

Baseball, 79
Bearcreek (Mont.), 109–10
Bell, Adam, 159
Berwind, *116*, 127

Berwind Mine, 115, 116, 127, 161
Bethlehem Steel, 43
Bonacci, Frank, 202
Bowers, Lamont Montgomery, 15, 26, 43, 49
Briggs, Howard, 207–8
Brissenden, Paul F., 111
"Brophy's" Mine, 109
Brotherhood of Locomotive Firemen and Enginemen, 151, 155
Brotherhood of Railroad Trainmen, 151, 155
Brotherhood of Railway and Steamship Clerks, 210
Burkhardt, John, 139
Bush, George W., 213

Cameron Mine, 113, 126, 131, 161, 302–3(n21)
Camp & Plant (journal), 75–76
Canada, 46, 219
Canadian Industrial Disputes Investigation Act, 47
Canon District, xi, 64, 124–25, 225
Carnegie Steel, 107
Castaneda, Jose, 96

317

INDEX

CF&I. *See* Colorado Fuel and Iron Company
CF&I *Blast* (journal), 78
CF&I *Industrial Bulletin*, 78, 79–80, 117, 283(n57)
Chinese, 90
Chorley, Kenneth, 165, 169
CIO: organizing for, 198–99; steel industry and, 194–95
Civil rights, workers and, 101–3
Clinton, Bill, 214
Coal camps: living and working conditions under, 73–74, 109–10; Rockefeller Plan and, 64–65, 67; welfare capitalism at, 75–76; YMCA at, 76–78
Coal Creek mine, 161
Coal mines/fields, *xii–xiii*, 89, 93, 179, 287(n41), 303–4(n21); control over work, 113–14, 122–23, 295(n21); Depression impacts, 178, 184–85; districts, 224–25; employee representatives, 239–53(table); ethnic makeup in, 89–90; safety, 118–20; strikes, 14–15, 159–60, 162–63, 168–70; unions, 172–73
Colorado and Wyoming Railroad, 18, 151, 155
Colorado Federation of Labor, and Rocky Mountain Fuel Company, 175–76
Colorado Fuel and Iron Company (CF&I), xi, 11, 25, 48, 49, 66, 71, 89, 108, 175, 182, 267(nn3, 5), 303–4(n21); bankruptcy, xiv, 302(n95); during Depression, 176–78, 184–85; eight-hour workday, 139–43; employee representatives, 44, 67–69, 85; and ERO, Inc., 202–3; labor relations, 5–6, 48, 185–86, 207–8; living conditions, 109–10; mine safety, 119–20; and NLRB, 200–201; and opposition to plan, 111–12; race and employment conditions at, 97–101; regularity of work in, 94–95; steel mill safety, 145–47; strikes at, 133–34, 135, 149–52, 162–63; UMWA, 127–29, 130–33, 179–80, 193, 208; wages, 124–26, 143–45, 160–61; welfare capitalism in, 74–79, 186–90
Colorado Fuel and Iron Corporation, 185
Colorado Industrial Commission, 70, 71, 117, 126, 161, 292(n51); Depression era impacts, 178–79; dispute resolution, 231–32; labor organization and, 149–50; on 1919 strike, 132–33; and 1927 strike, 163, 164, 169, 170–72; on safety issues, 118–19, 147
Colorado Industrial Plan. *See* Rockefeller Plan
Colorado National Guard, 16, 168, 293(n81)
Colorado Supreme Court, 22
Columbine Mine, massacre at, 159–60, 167, 168
Communications, Energy and Paperworkers Union, 219
Company unions, 209–10, 212, 268(n3), 309–10(n24); ban on, 215–16. *See also* Employee representation plans
Corwin, Richard W., 75
Creel, George, 23
Crested Butte, 178
Cuatro, 118
Curtis, Fosdick, and Belknap, 59, 94, 186

Democracy, industrial, xv–xvi, 6, 54–55, 217–18
Densmore, Warren, 80, 146
Depression: employee representatives during, 190–91; impacts of, 176–79, 184–85
De Santis, Dan, 33
Diamond, Andrew J., 79, 87, 137, 169, 181–82, 187; as employee representative, 85–86, 88, 190–91, 199, 200, 201, 307(n76); union opposition, 195, 202
Dickerson, Roy, 121
Dispute resolution, 113, 138; under employee representation plans, 103, 190–91, 228–32
District conferences, 225–26

INDEX

Dow, William, 101–2
Ducic, Paul, 198
Dunlop Commission, 215–17
Du Pont decision, 214

Eastern Europeans, 89, 90
Eight-hour workday, 236, 295(n21); negotiation for, 139–43
Elections: of employee representatives, 65, 67, 223–24; Minnequa Works, 201–2
Electromation, Inc., 213–14
Embree, A. S., 162
Employee Free Choice Act, 216
Employee representation plans (ERPs), xiv, 1, 43, 65, 71, 75, 96, 106, 133, 194, 209, 210–11, 217, 304(n39), 311(n42); criticism of, 8–9; during Depression, 190–91; development of, 26–27; impacts of, 38–39, 83–84; management responses to, 49–50; NLRA and, 62, 215–16; opposition to, 63–64; patterned after Rockefeller Plan, 82–83; popularity of, 2–3; Rockefeller's support for, 57–58, 182; during World War I, 81–82. *See also* Rockefeller Plan
Employee Representative Organization (ERO; ERO, Inc.), 201, 202–3
Employee representatives, 7, 34, 123, 169, 199, 286(n26), 292(n50), 306(n61); African Americans as, 101–2; class of work of, 92–93; at CF&I, 5–6, 44, 67–69; at coal mines, 239–53(table); Andrew Diamond as, 85–86, 88; dispute resolution, 117–21; election of, 65, 67; ethnic and racial issues, 103–4; and *Industrial Bulletin*, 79–80; on labor relations, 9–10; at Minnequa Works, 190–92, 197(table), 239–66(table); organization of, 201, 202; race and class of, 92, 96–97; and Rockefeller Plan, 35–36, 56–57, 72–73, 221–24; role of, 4–5; steelworkers, 138–39, 153–57; wage issues, 124–26
Engle mine, 122

English immigrants, 89
ERO. *See* Employee Representative Organization
ERP. *See* Employee representation plan
Ethnicity, 11, 104, 300(n40); of employee representatives, 239–66(table), 286(n26); of miners, 89–92
Eugenics, 105

Farrar, Fred, 21
Filler, Mike, 198
Fink, Walter, 20
Fitch, John, 127
Fitzpatrick, John, 149
Ford, Henry, 22
Fosdick, Harry Emerson, 147
Foster, William Z., 149, 152
Frederick Mine, 109, 302–3(n21)
Fremont County, 164, 225
Fremont Mine, 161
Friar, Merle, 138

Gary, Elbert, 7
Gates, Frederick, 49
Germans, 90
Germany, 219
Gilbert, William, 115
Glad, Mike, 132
Gold, Laurence E., 214–15
Gompers, Samuel, 7–8
Googhegan, Thomas, 213
Gould, Jay, 18
Great Coalfield War (1913–1914 strike), 14–15, 107; events of, 16–18; Ludlow Massacre, 19–20
Great Western Sugar Company, 99–100
Greeks, 90, 104
Green, William, 29
Greene, Jerome D., 46
Grievances, 113, 138; under employee representation plans, 103, 190–91, 228–32
Griffiths, David, 26–27, 36, 44, 67, 104, 114–15, 123
Gumbel, Robert, 189

319

Hair, R. L., 118–19
Hall, Henry, 43–44
Hapgood, Powers, 88, 109–10
Hayes, Frank, 21, 120
Health care, 75
Hefferly, Frank, 173, 198, 199, 307(n83)
Hefferly, Fred, 198, 199
Heydt, Charles, 44
Hiatt, Jonathan P., 214–15
Hicks, Clarence, 72, 76, 237, 238
Hispanics, 90–91, 96. *See also* Mexicans; Mexican Americans
Homestead Lockout, 107
Hood, Willis, 32–33
Hudspeth, Claude, 112
Huerfano County, 126, 164, 225
Hughes, Charles Evans, 210

Ideal Mine, 95
Immigrants: as coal miners, 89–90; IWW strike, 166–67
Industrial Commission (Colorado), 70, 71
Industrial Conference, 6, 22–23, 53, 58
Industrialization, 41–42
Industrial Relations Commission, 14, 21
Industrial Relations Counselors, Inc., 6, 42
Industrial Workers of the World (IWW), 12, 121, 300(n40); Colorado Industrial Commission on, 171–72; labor organization and, 172–73; and Rockefeller Plan, 161–62; strikes, 134, 162–67, 169; on strike violence, 159–60; vs. UMWA, 173–76
International Labor Organization, 135
International Union of Mine, Mill, and Smelter Workers, 201
Irwin, James, 191, 200, 201, 307(n76)
Italians, 90, 95, 104
IWW. *See* Industrial Workers of the World

Jacksonville Agreement, 161, 163
Japanese, 90
Johnson, George, 132
Joint Committee on Cooperation and Conciliation, 144
Joint Committee on Cooperation, Conciliation, and Wages, 144, 156
Joint Committee on Education and Recreation, 120–21, 227
Joint Committee on Industrial Cooperation, Conciliation (and Wages), 145, 226, 291(n20); dispute resolution, 230–32
Joint Committee on Safety and Accidents, 118–19, 146, 226–27
Joint Committee on Sanitation, Health and Housing, 120, 227
Joint Committees: dispute resolution, 117–21; executives and, 232–33; in Rockefeller Plan, 224–27
Jones, James, 96
Jones, Johnnie, 149
Jones, Mark M., 137
Jones, Mary "Mother," 18, 28, 29, 45, 61–62
Jurich, Frank, 162

Kaiser, C. A., 115
King, William Lyon Mackenzie, xiv, 6, 13, 15, *31*, 39, 41; employee representation plans, 43, 110; on labor relations, 11, 29, 30, 46–48, 49–51, 53–54, 67, 72; and John D. Rockefeller, Jr., 24–25, 30, 45, 56, 275(n9); and Rockefeller Plan, 62, 105
Ku Klux Klan, 100

Labor organization, xv, 53, 82, 130, 131, 211–12, 295(n21); and Colorado Industrial Commission, 149–50; IWW, 172–73; NLRB support, 213–14; race and, 104–5; steel industry, 194–95
Labor question, xvi, 41–42
Labor reform, 6, 55, 217–18
Labor relations, xiv, 210; CF&I, 5–6, 207–8, 281(n32); collective bargaining, 10, 68, 309(n13); employee representatives and, 9–10; ERPs and, 38–39, 55–59; Mackenzie King on, 46–48, 49–50, 53–54; power relationships, 48–49;

INDEX

John D. Rockefeller, Jr., 6–7, 22–23, 24–25, 28–29, 39, 41–42, 182–83, 204–5, 278(n62); Charles M. Schwab on, 42–43; steelworkers, 136–37
Las Animas County, 164, 225
Latchem, E. W., 162
Lawson, John, 21–22, 77
League of Nations, 136
Lee, A. H., 140
Lee, Ivy, 19, 23, 24
Lester Mine, 121–22
Lewis, John L., 52, 173
Liberal Party (Canada), 46
Lichty, Arthur H., 45
Living conditions: for miners, 109–10; under Rockefeller Plan, 73–74, 109, 235–36
Lockouts, 47
Loeffler, Joseph, 139
Low, Seth, 25
Low Commission, 25, 26
Ludlow Massacre, xiv, 2, 16, 17, 270(n4), 271(n19); impacts of, 10–11, 70; memorial dedication, 13–14; press coverage of, 19–20; publicity over, 20–21, 23; and John D. Rockefeller, Jr., 15, 18–19, 23–24, 39, 43

McBrayer, J. L., 167–68
McGee, Thomas, 79–80
McMahan, Ronald L., 162
Madden, Warren, 200–201
Management, 54; employee representatives and, 67–68, 123; and joint committees, 117–21; and labor, 47–48, 114; power relationships, 69–70, 80–81; under Rockefeller Plan, 71–72; and steelworkers, 138–39, 155–57, 185–86
Managers, 48; and employee representatives, 4–5; Rockefeller Plan, 35–36
Martinez, Valentino, 97, 98
Matteson, B. J., 102, 115–16, 117, 138, 156
Maxwell, W. A., 205
Mayo, Elton, 86, 108, 183, 283(n68); on power relationships, 80–81

Merritt, John, 69
Mexican Americans, 91, 92, 105, 106, 285(n13); during Depression, 177–78; employment conditions, 95, 96, 97–101, 115–16; and IWW, 166–67; miners, 88, 89, 287(n42); steelworkers, 155, 195
Mexicans, 91, 106, 155, 285(n13); during Depression, 177–78; employment conditions, 95, 96, 97–101, 115–16; and IWW, 166–67, 300(n40); miners, 88, 89–90; and Rockefeller Plan, 88, 92, 121; SWOC, 195, 196, 198–200; and UMWA, 104–5
Miners, 5, 107; Depression impacts, 177–78; employee representatives and, 92–93; ethnic makeup of, 89–92; living conditions, 109–10; 1927 strike, 162–63, 164–68; regularity of work for, 93–95; Rockefeller mine tour, 32–33; Rockefeller Plan, 35–36, 111–12; suspension and dismissal of, 121–24; unionization, 172–76, 179; wage cuts, 160–61; wages, 124–26, 286(n27); working conditions, 113–14, 287(n42)
Mines. *See* Coal mines/fields
Mining camps, 109–10, 120; welfare programs at, 75–78; YMCAs at, 76–78
Minnequa Works, xi, 12, 56, 95, 97, 108, 124, 169, 181; eight-hour workday, 140, 141; employee representatives at, 85, 106, 197(table), 239–66(table); independent union formation at, 194–95; Rockefeller Plan at, 33, 64, 180, 183–84, 195–96; safety in, 145–46; steelworkers rights at, 138–39; strikes at, 136, 137, 149–52; union elections in, 201–2, 203–4
Mondragon, Cesario, 119
Murders, UMWA, 21–22
Murphy, Starr, 25–26, 29, 54, 129

National Association of Iron and Steel Workers, 149, 150–51
National Association of Manufacturers, 83

INDEX

National Committee for Organizing Iron and Steel Workers, 136, 149
National Industrial Recovery Act (NIRA), 8, 75, 179, 193, 211
National Labor Relations Act (NLRA), 3, 8, 12, 27, 75, 193, 195, 198, 200, 210, 213, 306(n61), 309–10(nn13, 24); on employee participation programs, 215–16; on labor organization, 211–12; and sham unions, 62–63
National Labor Relations Board (NLRB), 193, 212; Minnequa Works elections, 201–2, 203–4; and Rockefeller Plan, 68, 196; and SWOC, 198, 199, 200–201; and unions, 213–14
National Labor Relations Board v. Newport News Shipbuilding & Dry Dock Co., 212
National War Labor Board (NWLB), 81–82, 129
Nimmo, John, 22
NIRA. *See* National Industrial Recovery Act
NLRA. *See* National Labor Relations Act
NLRB. *See* National Labor Relations Board
NLRB vs. Jones and Laughlin Steel Corporation, 195
NWLB. *See* National War Labor Board

Osgood, John C., 18
Ostrander, Earl, 147, *148*, 296(n51)

Palmer, Frank L., 164
Parks, F. E., 149
Payne, Thomas, 125
Peters, Marco, 104
Petrucci, Mary, 21
Pictou Mine, 95, 111, 303–4(n21)
Pinchot, Gifford, 22
Pittsburgh Plus, 143
Plute, George, 138
Pogliano, Felix, 164
Poles, 90
Power relationships, 48–49; under Rockefeller Plan, 69–70
Primero, 118
Princeton University, 42
Promotions, steelworks, 155–57, 202
Propaganda, 23; radical, 121–24
Pueblo, xiv, 75, 135; racism in, 100–101; steel strike in, 149–52; steel works in, xi, 184, 267(nn3, 5, 6)

Quigg, Louis F., 138, 183–84, 185, 186, 190, 191–92, 203

Race, 11, 88, 103, 188; anti-strike efforts, 167–68; and employee representatives, 92–93, 239–66(table), 286(n26); and employment conditions, 94–95, 97–101, 102–3; Hispanics, 90–92; and labor organization, 104–5; and 1927 strike, 166–67; steelworkers, 95–96. *See also* Ethnicity
Racism, 100–101
Railway Labor Act, 210, 211
Ramey, 127
Ratkovitch, Max, 122–23
Reed, John, 105
Roche, Josephine, 160, 174, 175, 178
Rockefeller, Abby, 13, 56, 147
Rockefeller, John D., Jr., xiv, xvi–xvii, 5, 13, 26, 40, 76, 105, 181, 187, 272(n47), 274(n77); anti-unionism of, 29–30; Colorado mine tour, *31–33*; and Depression-era wage cuts, 178–79; industrial relations philosophy of, 58–59; and Mackenzie King, 24–25, 45–46, 47–48, 275(n9); labor relations, 6–7, 22–23, 28–29, 41–42, 110, 136, 182, 204–5, 208, 210, 278(n62); and Ludlow Massacre, 10–11, 14–15, 18–19, 23–24, 39; and Rockefeller Plan, xvi, 2, 33–35, 37–38, 43–45, 55–59, 136; on unions, 22–23, 51–53
Rockefeller, John D., Sr., xvii, 18, 42
Rockefeller, John D., III, on Rockefeller Plan, 182–83
Rockefeller, Nelson, 187
Rockefeller family, xiv, 14, 19, 42

322

INDEX

Rockefeller Foundation, 25
Rockefeller Plan (Colorado Industrial Plan), xiv, xv–xvi, 1–2, 5, 9, 19, 75, 88, 108, 176, 208, 215, 280(n14), 306(n63); administration of, 183–84; adoption of, 29, 32, 33–36; criticism of, 40, 61–62, 112–13, 114–15; design of, 64–67; development of, 25–27; dispute resolution, 117–21; employee representatives under, 67–68, 85–86, 92–93; employee rights under, 72–73; ERPs patterned after, 82–83; evaluation of, 182–83; failures of, 199–200; IWW and, 161–62; at Minnequa Works, 180, 195; negotiations under, 153–55; NLRB on, 200–201; 1927 strike, 169–71; Gompers on, 7–8; implementation of, 43–45; and independent unions, 71–72; King's goals for, 50–51; limits to, 3–4; living and working conditions under, 73–74; Minnequa Works, 195–96; opposition to, 111–12, 131–32, 133, 134, 161–62, 170–71, 177; power relationships under, 48–49, 69–70, 80–81; Rockefeller's goals for, 37–38, 51–52, 136, 278(n62); Rockefeller's interests in, 43–44, 55–59; Russell Sage Foundation study of, 112–13; social order in, 105–6; steelworkers, 152–57, 282(n40); text of, 221–38; UMWA and, 126–29, 179; wages issues, 124–26, 145, 160–61; workers' power under, 48–49; worker response to, 115–16; and YMCA, 188–89
Rockvale, 123, 161, 302–3(n21)
Rocky Mountain Fuel Company, 160; and CF&I, 178–79; and UMWA, 173, 174, 175–76
Roeder, Arthur, 176, 178, 179–80, 185, 187, 188, 191
Roosevelt, Franklin D., 219
Roosevelt, Theodore, 23
Rouse (Colo.), 101–2
Rupp, G. H., 177
Russell Sage Foundation, 112–13

Sacco, Nicolo, 162
Safety: mine, 118, 119–20; steel mill, 145–47
Sandberg, Carl, "Ivy L. Lee—Paid Liar," 23
Sanitation, 75
Schwab, Charles M., 42–43
Scottish immigrants, 89
Segregation, at YMCAs, 101–2
Segundo Mine, 162
Selekman, Ben M., 40, 69, 73, 123, 140; on CF&I operations, 92–93, 112, 132, 139, 144, 146, 297(n69); on Rockefeller Plan, 114–15
Seniority, promotions and, 155–57, 203
Shafer, Harrington, 146
Shepherd, John, 167–68
Sinclair, Upton, 23, 272(n47)
Skidmore, W. E., 237, 238
Slichter, Sumner, 83
Snyder, Frank, 16, *17*
Social order, Rockefeller Plan as, 105–6
Sociological Department (CF&I), 75–76
Sopris, 119
Spanish Americans, as term, 90–92
Starkville Mine, 111
Steel industry, 184, 185, 186, 267(n3); eight-hour workday, 139–43; labor organization, 194–95; in Pueblo, xi, 89; Rockefeller Plan negotiations, 153–57; safety in, 145–47; strikes in, 149–52, 296(n54), 297(n69); trade unions, 107–8; wages, 143–45
Steelworkers, 185; eight-hour workday, 139–43; employee representatives and, 11–12, 96–97, 155; employment conditions at, 95–96, 97–98; labor relations, 136–37; and management, 185–86; on Rockefeller Plan, 152–53, 282(n40); safety, 145–47; strikes, 135, 149–52, 296(n54), 297(n69); trade unions, 107–8; union organization, 198–99, 203–4; wages of, 143–45; working conditions, 153–54; YMCA and, 187–88
Steel Workers Organizing Committee (SWOC), 12, 194–95, 202, 203,

323

304(n35), 307(n83); formation of, 196, 198–200; and NLRB, 200–201
Stewart, A. T., 202
Stout, David, 161
Strikes, 9, 47, 103, 104, 126; CF&I and, 25, 122, 127, 128, 133–34, 135; Great Coalfield War, 14–15, 21; and new hires, 89, 90; 1919, 129–32; 1927 coal mine, 162–72; steel industry, 136–37, 149–52, 296(n54), 297(n69); violence, 159–60
Sugar industry, 99–100
Sunrise (Wyo.), xi, 104, 201
Sunrise District, 225, 307(n78)
Swedes, 90
SWOC. *See* Steel Workers Organizing Committee

Tabasco, 127
T&NO. *See* Texas and New Orleans Railroad
Taft, William Howard, 81
TEAM. *See* Teamwork for Employees and Managers Act
Teamwork for Employees and Managers Act (TEAM), 214–15, 216
Tercio, 118
Texas and New Orleans Railroad (T&NO), 210–11
Texas and New Orleans Railroad Company v. Brotherhood of Railway and Steamship Clerks, 210
Thompson, Carl, 187
Tikas, Louis, 16
Toller Mine, 122
Towson, Charles R., 77
Trinidad District, *xii–xiii,* 64, 90, 128, 224–25
Turner, W. D., 153

UMWA. *See* United Mine Workers of America
Unions (trade unions), 107, 130, 179, 208, 219–20, 307(n78); coal miners, 172–73; employee representation plans and, 3, 8, 84, 106, 311(n42); independent, 7, 194–95; Minnequa Works, 201–2; NLRA and, 211–12; NLRB and, 213–14; King on, 50–51, 53–54; Rockefeller Plan and, xv, 71–72; Rockefeller on, 43, 51–53; sham, 62–63; steel industry, 149–52, 194–95, 196, 198–99, 203–4; UMWA vs. IWW, 173–76; World War I, 81–82
United Mine Workers of America (UMWA), 2, 10, 25, 51, 52, 84, 107, 120, 124, 161, 164; CF&I and, 130–33, 179–80, 193; Great Coalfield War, 16–17; impacts of, 109–10; Industrial Conference, 22–23; vs. IWW, 173–76; and Ludlow Massacre, 13–14, 17, 20; opposition to Rockefeller Plan, 111–12, 126–29; propaganda, 22–23; race and, 104–5; and John D. Rockefeller, Jr., 26, 29, 30; and Rocky Mountain Fuel Company, 175–76; strikes, 131–32; violence, 21–22
U.S. Chamber of Commerce, War Emergency and Reconstruction Conference, 57–58
U.S. Commission on the Future of Worker-Management Relations (Dunlop Commission), 215–17
U.S. Department of Labor, 65
U.S. Industrial Relations Commission, 14, 23, 24, 45, 114; Rockefeller's testimony, 28, 43, 51
U.S. Steel Corporation, 143; eight-hour workday, 140, 142; unions, 107–8, 149
United Steel Workers of America (USWA), 203, 204
U.S. Supreme Court, labor decisions, 210–11, 212
USWA. *See* United Steel Workers of America

Vacations, for steelworkers, 152–53
Valdez (Colo.), 109, 177, 178
Van Dyke, Frank, 143
Van Kleeck, Mary, 40; on CF&I operations, 112–13, 114–15, 120, 132

INDEX

Vanzetti, Bartolomeo, 162
Victor American camps, 110
Villa, Jose, 93
Vincent, Merle, 164
Violence, 20; Great Coalfield War, 16–17, 20; Ludlow Massacre, 270(n4), 271n(19); strikes, 159–60; UMWA, 21–22
Vosnica, Joe, 170–71

Wages, 163, 194; CF&I cuts in, 160–61, 178–79; disputes over, 124–26; and eight-hour workday, 140, 141, 142; and employee representative plans, 194, 236–37, 307(n76); 1927 strike, 164, 166, 169–70; steelworkers, 143–45, 186
Wagner, Robert, 8, 62, 211–12
Walsenburg District, *xii–xiii*, 64, 225; employee representatives in, 92–93; strikes, 128, 166
Walsen Mine, 167
Walsh, Frank, 14, 22, 28
War Emergency and Reconstruction Conference, 57–58
Weir, James, 122–23
Weitzel, Elmer, 44, 52, 99, 111, 124, 128
Welborn, Jesse, 23, 26, 27, 48, 49, 94, 95, 106, 111, 113, 127, 128–29, 143, 147, 170; on eight-hour workday, 140–42; on 1919 strike, 149, 150, 151, 152; on 1927 strike, 165–66; on Rockefeller Plan, 50, 65, 67, 72; and steelworkers, 156–57; on strike, 131–32
Welfare capitalism, 110; at CF&I, 74–79, 189–90

Welsh immigrants, 89
West, George P., 24, 114
Western Federation of Miners, 107
Western Slope District (Western District), *xiii*, 64, 225
Western States Steel Products Union, 195
White, John P., 29
Williams, Don, 156
Wilson, Woodrow, 20, 26, 130, 135, 136; Great Coalfield War, 16, 25; Industrial Conference, 6, 53
Woll, Matthew, 179
Workers: on civil rights, 101–3; and employee representatives, 92–93; employment conditions, 93–95, 96–101, 281(n32); industrial democracy and, 217–18; and labor relations, 46, 48–49; response to plan, 115–16; unions and, 51–52, 213–14, 219; and YMCA, 76–78, 79
Working conditions, 237; power relationships and, 48–49; under Rockefeller Plan, 73–74
Workplace Relations Act (Australia), 219
World War I, 23, 55, 130; steelworkers, 140–41; unionism, 81–82

YMCA. *See* Young Men's Christian Association
Young, Adam, 166
Young Men's Christian Association (YMCA), 109, 139, 304(nn27, 35); closing of, 186–89; at mining camps, 76–77; participation in, 77–78, 79; racism and segregation in, 100–102